Around the world, corporate scandal – and full-scale collapse – has caught the headlines in a spectacular way, and investors avidly search for scapegoats. We all express surprise at such catastrophes – yet extinction is an inherent fact of life, and failure comes calling at the door of companies both gigantic and small.

Over 17,000 companies will go bust this year in the UK alone. But is this a bad thing? And if so why does the US, with its hugely dynamic economy, see more than 10 per cent of companies disappear each year?

Paul Ormerod was for several years Head of the Economic Assessment Unit at *The Economist*. He was Director of Economics at the Henley Centre for Forecasting from 1982 to 1992, and has been a visiting Professor of Economics at London and Manchester. He has published widely in academic journals, lectures around the world, and is currently a director of Volterra Consulting. He is the author of *Butterfly Economics* – 'a fascinating and entertaining introduction to the economics of the 21st century' (*New Scientist*) – and *The Death of Economics*, which the *Guardian* found 'controversial and impassioned' while the *Spectator* declared that it 'should be read by every educated person.'

Further praise for *Why Most Things Fail*:

'There is a powerful message here.' Hamish McRae, *Independent*

'Interesting and entertaining ... The scale and breadth of Ormerod's analysis deserves commendation.' Adrian Woolfson, *Nature*

Why Most Things Fail
Evolution, Extinction and Economics

PAUL ORMEROD

faber and faber

First published in 2005
by Faber and Faber Limited
3 Queen Square London WC1N 3AU

This paperback edition first published in 2006

Photoset by RefineCatch Limited, Bungay, Suffolk
Printed in England by Mackays of Chatham Ltd

A CIP record for this book
is available from the British Library

ISBN 0–571–22013–4

10 9 8 7 6 5 4 3 2 1

Contents

Preface

An argument I first made ten years ago in my book *The Death of Economics* noted that conventional economics views the economy and society as machines, whose behaviour, no matter how complicated, is ultimately predictable and controllable. On the contrary, however, human society is more like a living organism.

My previous book, *Butterfly Economics*, which was first published in Britain in 1998 and has subsequently appeared in many languages around the world, developed this theme. I analyzed a wide and seemingly disparate range of economic and social questions, seeing them as analogous to living creatures whose behaviour can only be understood by looking at the complex interactions of their individual parts.

The idea that economics should look to biology for intellectual inspiration is a long-held and distinguished one. Alfred Marshall, who founded the faculty of economics at Cambridge University in around 1900, was the first major scholar to articulate this view. Later, Friedrich Hayek, perhaps the most innovative social scientist of the twentieth century and a thinker many years ahead of his time, stressed the importance of dynamic, evolutionary change in the workings of human social and economic systems, while Joseph Schumpeter of Harvard wrote famously of the 'gales of creative destruction' which he regarded as the defining principle of the market-oriented capitalist economies.

These ideas have never really been absorbed into the economics mainstream. Stability, order and equilibrium continue to be emphasized when the real world is characterized by constant change, evolution and disequilibrium. Part of the reason for the failure of economics to move in this direction is sheer intellectual inertia. More forgivingly, the tools required for the systematic analysis of systems of this kind have only recently become available. *Butterfly Economics* gave a number of practical examples where the new kinds of analysis required to understand such complex systems provide better empirical explanations of how the world works than does the machine-like views of economic orthodoxy.

In this book, I address what is probably *the* most fundamental feature of both biological and human social and economic systems: failure. Species fail and become extinct, brands fail, companies fail, public policies fail. Fortunately, not everything fails at the same time. And sometimes we can see examples of success which persist for long periods of time. The Roman Empire, for example, lasted for many centuries. But to understand success we must first understand the pervasive existence of failure. The documentation of failure, the identification of subtle patterns amongst the apparent disorder of failure, and analyzing why failure arises are the main themes of this book.

I am grateful to a number of people for encouragement and discussions which have helped to develop the ideas of this book, and in particular to Rich Colbaugh, Bridget Rosewell and Bob Rowthorn. Julian Loose of Faber and Faber has once again proved to be a very helpful and inspirational editor.

PAUL ORMEROD

London, Wiltshire and New Mexico, August 2004

Introduction

Failure is all around us. Failure is pervasive. Failure is everywhere, across time, across place and across different aspects of life. Ninety-nine point nine nine per cent of all biological species which have ever existed are now extinct. Failure in this context is measured over hundreds of millions of years. On a dramatically shorter timescale, more than 10 per cent of all the companies in America disappear each year. Large and small, from corporate giants to the tiniest one-person business, they fail.

Over the last fifty or sixty years, western governments have intervened to try to improve the social and economic life of their countries on a scale unimaginable to previous generations. Yet social and economic problems persist. Policies fail.

Two examples will suffice. Despite decades of governmental activity to promote integration, residential areas in many western countries remain strongly segregated along racial lines. For many years, the powers that be have also tried to increase social mobility, to ensure that the children of the poor have as many opportunities to better themselves as the children of the rich. Yet the evidence shows that social mobility, far from increasing, has actually fallen in recent years.

From biological species to companies to government policies, there appears to be an Iron Law of Failure, which is extremely difficult to break. Yet the existence of failure is one of the great unmentionables. Within economics, we will look in vain for any satisfactory account of why firms fail. Business gurus eulogize contemporary success, conveniently ignoring the fact that many of these firms often fall on harder times soon after receiving their acclaim. Enron, for example, was praised to the skies for its dynamism and innovative thinking right up to the point when it became the epitome of corporate greed and maladministration.

In the more abstract world of economic theory, we can find a great deal of material, much of it irrelevant or even misleading, on what firms should do in order to succeed. But there is little, if anything, about why firms fail.

Charles Darwin's brilliant theory of evolution, expounded in the middle of the nineteenth century, explains not why species fail, but why they succeed. In the Darwinian theory of the process of evolution, species gradually become better adapted to their immediate environment, become fitter for survival.

In spite of this, nemesis eventually claims them, and species become extinct. The survival of the fittest comes up against the Iron Law of Failure. How can it be that even the fittest, honed and toned to compete in the struggle for survival, fall by the wayside and disappear for ever? The endemic failure at the level of the individual species confronts us with this paradox. Darwin wrote almost 150 years ago, but it is only very recently indeed that biological theorists have begun to analyze systematically the evidence of failure at the species level.

The Iron Law of Failure appears to extend from the world of biology into human activities, into social and economic organizations. The precise mathematical relationship which describes the link between the frequency and size of the extinction of companies, for example, is virtually identical to that which describes the extinction of biological species in the fossil record. Only the timescales differ.*

The book addresses some intellectually demanding topics. Many of the results that are referred to, particularly in the second half of the book, are based upon mathematical models of aspects of both society and the economy. But I can reassure readers from the outset that this book is written entirely in English, aided to some extent by charts and graphs. Those who crave mathematical

---

* I have published this finding in *Physica A*, the world's leading journal of statistical physics.

detail can readily satisfy their urges in the references listed at the end of the book.

The main theme of this book is to develop a general explanation of the pervasive nature of failure in the world of human societies and economies. Though there are striking parallels between the social and economic world and the world of biology there is, however, a fundamental difference between the two: the process of evolution in biological species cannot be planned. Species cannot act with the intent of increasing their fitness to survive. In contrast, in human society, individuals, firms and governments all strive consciously to devise successful strategies for survival. They adapt these strategies over time and alter their plans as circumstances change. Yet, despite this apparent contrast, eventually, in both biological evolution and human social and economic activity, failure strikes.

A second theme of this book is to understand this seeming paradox. How can it be that not just failure, but the patterns of failure, are so similar in biology and human organization when there is such a sharp contrast between the abilities to act with the conscious intent of improving one's prospects for survival?

The third theme, developed in particular towards the end of the book, is that failure can be highly beneficial. In the real world in which strategies evolve and which is itself the outcome of a dynamic process of change, failure at the level of the individual component part can, paradoxically, enhance the fitness of the system as a whole.

The book's content is firmly grounded in reality. Too much work in the social sciences, whether it is the dense mathematics of much of economics or the tortuous prose of a great deal of sociology, is purely theoretical. Throughout the book, I compare the theory with the evidence of what we actually observe. This is the basis on which the natural sciences have achieved such tremendous success in understanding the physical world over the past few centuries. To be of any value, theories must be confronted with reality.

1 The Edwardian Explosion

The period from around 1880 to 1910 saw the emergence of radically different ways of organizing and carrying out economic activity. The consequences both for economic well-being and the wider sphere of political economy were dramatic, so we will begin by exploring the developments during this period in some detail.

The eminent biologist Stephen Jay Gould coined the phrase 'the Cambrian explosion' for the period some 550 million years ago when, suddenly, dramatic new life forms surged into being. After an immense length of time during which life had existed in only its simplest forms, far more complicated creatures came into existence. Most prospered for a while and then failed. But the legacy was the path of evolution which has led eventually to humanity.

Similarly, in the economic world, the decades around 1900 saw massive companies emerge for the first time, bringing entirely new management problems in terms of co-ordinating and organizing the operations of these vast entities. In British social history, this is known as the Edwardian period, after Queen Victoria's son who himself reigned in the opening decade of the twentieth century, a period when the British Empire dominated the world. So perhaps I might be permitted a trace of nostalgia in describing the events of this period, so important to the future development of capitalism, as the 'Edwardian explosion'.

During these few decades, we can see forms of organizing economic activity fall by the wayside as firms struggled to understand and adapt to the rapidly changing environment. Yet, at the same time, the survivors from this turbulent age were successful on a scale entirely without precedent. The modern world of huge multinational companies, so familiar to us now, was essentially created during this period. Globalization is a hot topic in the early twenty-first century, but its foundations were laid a century before.

The single most useful and productive legal invention in the past few centuries has been that of the commercial firm. Individuals have banded together and pooled their resources in the pursuit of business since time immemorial, but the massive economic expansion of the past two hundred years is based on the modern concept of the company. Financed by outside shareholders and facing limited liability, this new way of organizing the production of goods and services has transformed the world. The firm is the *Tyrannosaurus rex* of economic activity, a hugely successful species that sweeps all before it.

We can identify two features of the company which make it qualitatively different to all previously existing ways of conducting business. Each is important in its own right, but combined their joint impact is greater than the sum of the individual components. Both had been invented prior to the final quarter of the nineteenth century. But it was during this period that the overall conditions became right for a dramatic transformation of the economic environment based on them.

The first feature is the idea of attracting outside investors into the venture. By itself, this is not particularly new. Wealthy individuals have always been willing to put money into the ideas of talented and resourceful people known to them. Much of the world's great art and music, for example, was financed by private donations from the rich. What is different about the modern firm is that the investment is essentially anonymous. Shareholders do not need to know personally the entrepreneurs in order to part with their money. Of course, with start-up companies or small firms looking for finance to expand, a prudent investor will insist on finding out a great deal about the individuals concerned, while a fund manager looking to move a big block of shares between, say, Microsoft and GM may well think quite hard about the key individuals on the boards of these giant companies. But there has been a massive expansion in the amount of information available to investors about what is on offer. Companies can and do solicit new funds from individuals who are completely unknown to them at the time, and this new form of organization increases dramatically the potential funding for any individual enterprise.

The second feature, or evolutionary step as we might think of it, is the invention of limited liability. Individuals need no longer risk personal bankruptcy when they organize a commercial venture. Indeed, one may feel that this particular quality has recently taken an evolutionary step too far. Managers, facing no personal risk whatsoever, reap spectacular rewards for failure – failure with other people's money. A form of corporate theft has been perpetrated in many cases.

But this latter is a very recent phenomenon, and the contribution of the concept of limited liability has been hugely positive. All business decisions involve risk. The degree of risk may vary enormously, but no one knows for certain what will happen once a decision is taken. The limit placed on liabilities facilitated an explosion of innovation and entrepreneurial activity. Individuals were released from the constraint of, quite literally, having to bet the family ranch on a new business venture.

Of course, out in the thickets of the commercial world, different species, different forms of corporate organization survive, each with its own niche. Some can be very successful. Goldman Sachs, for example, has been one of the most profitable, dynamic and innovative financial institutions in the world in recent decades. And for most of this time, it was an antiquated partnership, fashioned on the same organizational principles as those of the bankers who financed Europe's monarchs in medieval times. In both cases, the potential rewards were huge. On the downside, however, the entire personal wealth of each individual partner was, in principle, at risk every single day.

The dominant life form for more than a century, however, has been that of the limited liability company. Like the dinosaurs, this took time to reach its full evolutionary potential. The massive dinosaurs that ruled the world did not spring up entirely from nothing. In the same way, the concepts of anonymous outside investors and limited liability were not invented in the final quarter of the nineteenth century. But, suddenly, underpinned by these concepts, the conditions became right for a massive step forward in the evolution of firms. Companies grew stupendously, to sizes that were entirely without precedent in human history. At the turn of the nineteenth century, large corporations were being built on an enormous scale, mainly due to a massive wave of mergers and acquisitions. By the first decade of the twentieth century, for example, US Steel employed more than 200,000 workers, a number simply beyond the imagination of previous generations.

US Steel was admittedly by far the largest company in the world at that time. Its total assets in 1917, for example, were no less than $2,449 million. Translating this into modern prices is not straightforward because so many things have changed since then, but an approximation would be a value of some $400 billion. For comparison, the value of Microsoft is currently around $300 billion. So US Steel was big by any standards.

But many other American companies had assets of over $100 million, with eleven more companies in exactly the same industrial sector as US Steel – 'primary metal industries' in the dry jargon of economic statistics. The industry of 'transportation equipment' had been made up of locomotive and ship manufacturers until the beginning of the twentieth century, but as early as 1917 the largest firm in this sector was already the new Ford Motor Company, with assets of $165 million. In third place in this list was another familiar name, that of General Motors. Elsewhere in the economy, giant corporations had sprung into existence. The food sector, for example, was headed by Armour and Co. and by Swift and Co., each with assets of over $300 million. Both of these became extinct as independent firms in the 1970s and 1980s, respectively. Du Pont and Union Carbide were the largest producers of chemicals, and Standard Oil of New Jersey the biggest oil company, with assets of over $500 million.

The success of the large company, far more efficient and productive than anything that had gone before, was instrumental in consolidating the political success of capitalism, itself a relatively new form of economic life, which had evolved gradually from its feeble initial stirrings in the Europe of the sixteenth and seventeenth centuries.

Living standards had been improving gradually during the nineteenth century. There is a bitter and intense debate, seeming to stem as much from ideology as from objective scholarly dispute, about whether average living standards rose or fell in the early decades of the Industrial Revolution, up until around 1840. But all are agreed that from around that time life improved. The number of hours worked per week were reduced, health began to improve along with life expectancy as people could afford to buy more food and hence consume more calories, and more and more products appeared in the shops which came within the reach of ordinary people.

Nonetheless, life was undoubtedly still pretty grim for most people. Again, comparisons across such a long period of time are difficult to make, since the whole structure of the economy and the mix of goods and services which are now available have altered dramatically. Most of the purchases made today are of products which simply did not exist a century or more ago. Air travel is an obvious example, but the inventiveness of capitalism knows no bounds. As I write these words, I read in the newspaper of a German savaged almost to death by his pet Rottweiler. He was attempting to give it fresher breath by brushing its teeth. Confronted by doctors telling him he was lucky to be alive, the man moaned plaintively, 'I can't understand it, I used a special canine toothbrush.' We can be certain that special canine toothbrushes were not generally available in the shops of the late nineteenth century.

Despite the difficulties, economists and economic historians have made great efforts to compare living standards over time. As a broad generalization, we might say that the average person in western Europe in 1880 was as well off as, say, the typical Indonesian today. If anything, the European of that time was slightly worse off, but the comparison is not unreasonable. The threat of famine, persistent for millennia, was starting to fade into

5

memory and surplus cash was becoming available to spend on things other than the bare necessities of life. But work was long and hard, support from the state at times of personal hardship was weak or non-existent, and the environment in the cities was badly polluted, far more so than it is today.

If living standards had not improved any further, it is easy to believe that the appeal of a political platform dedicated to the revolutionary overthrow of capitalism would have grown. The stupendous cornucopia that capitalism was to unleash in the twentieth century in terms of material standards of living, health and life expectancy across the planet could scarcely be imagined. In any event, that was just speculation about the future rather than the bleak reality of actual experience.

However, the success of the newly evolved life form of the large company gave a huge boost to the quality of life in the period from 1880 to the First World War. The average standard of living in America rose by 50 per cent. In France and Germany the increases were 50 and 60 per cent respectively, and in Britain, the wealthiest of the major countries in 1880, incomes rose by a further 40 per cent. As a result, the attraction of a revolutionary change faded away. Apart from brief insurrections in the aftermath of the First World War, capitalism has remained unchallenged in the west throughout the twentieth century. Only in Russia in 1917 do we have an example of the overthrow of the system.

This period of immense political and economic significance for the future of both the west and the world in general also saw the rise of a new political species, namely the social democratic party. Despite the rhetoric, these parties were effectively committed, not to the elimination of capitalism, but to a modified version of the system. Their true nature was revealed for all to see by the First World War itself. A small minority urged the workers of Europe to resist the war. In contrast, the great majority of social democrats everywhere threw themselves with enthusiasm into their national struggles. A leading revolutionary, Rosa Luxemburg, attacked the role of social democratic parties in vituperative terms a few days before her assassination in the turbulent political atmosphere of

Germany of December 1918: 'In all previous revolutions, the combatants faced each other openly, class against class, programme against programme. In the current revolution, the troops protecting the old order are not intervening under the banner of the ruling class, but under the flag of the social democratic party.' But from the perspective of the people they represented, the social democratic leaders were acting in an entirely rational way. Capitalism was delivering the goods.

Returning to the purely economic aspect rather than the wider political economy of this period of dramatic change, our knowledge of the emergence of the massive company in the decades around 1900 has been expanded enormously by the work of Alfred Chandler, an economic historian at Harvard. His magnum opus, *Scale and Scope: the Dynamics of Industrial Capitalism*, charts in great detail the development of large firms in America, Britain and Germany, the three biggest economies of the world at the time.

The pace of innovation and evolution of economic organization was rapid, matched by a wave of extinctions as older, less efficient structures failed to survive in the new environment. Consumers today are quite accustomed to the idea of the prices of new products and services falling dramatically; for example, computer hardware, software and telecommunication products have all seen sharp falls in price combined with better quality products. Exactly the same phenomenon was observed in the cutting-edge industries of the late nineteenth century. Their products were distinctly less exotic than the electronic-based wave of today, but then, as now, new ideas revolutionized production. The German dye manufacturers Bayer, Hoechst and BASF were able to reduce the price of a new synthetic dye from 270 marks a kilo in 1869 to just 9 marks by 1886. Less efficient producers were simply swept aside.

The period saw a veritable explosion of different types of industrial structure. Managers and owners searched and explored a wide variety of strategies in the pursuit of success in the rapidly changing environment. Many were tried, only to be discarded rapidly. Large-scale investment in new machines and factories and rapid technological innovation in many industries in the 1870s and 1880s brought success for some and great benefits to the consumer. But success came at a price to the companies themselves. Competition intensified and prices fell. The example of synthetic dye is an extreme example of what happened in what was then a very high-tech industry. But even in well-established industries such as textiles and iron, innovation led to lower prices. And lower prices meant lower profits.

A widespread response, in both Europe and America, was to make formal agreements between companies, enforced by trade associations. Chandler notes that in the American hardware industry alone, which had many highly specialized product lines, no fewer than fifty trade associations managed the market for firms. These associations set quotas for output, fixed prices and allocated different regions to different companies. The intense level of competition and innovation threatened the very existence of many firms, and the initial response was to band together to manage and control this frightening new beast. Competition gave way to collaboration.

But the policy of attempting to manage competition through cartels organized by trade associations pretty quickly failed. Essentially, there was no effective mechanism of policing the agreements on prices and output. Each individual member of an industry cartel faced a strong incentive secretly to cut prices to gain business. And once one firm had broken ranks, the others were compelled to follow suit, if they could. The less efficient were forced out of business because they were unable to make profits at the new, lower levels of price, and profits were lower than they were for those who survived.

It was always tempting for an individual to break the cartel, with the eventual result that all participants in the agreement ended up worse off. This conflict between what we can think of as individual and collective rationality is one which we will encounter again in Chapter 5 when we discuss game theory. It was in the collective interests of firms to maintain the cartel, but individual firms often came to believe that they themselves would be better off by breaking it.

8

This failure of the structure within individual industries was soon reinforced in the US by powerful external pressure. The Sherman Antitrust Act of 1890 not only made such combinations illegal but provided the federal government with the authority to enforce this through the courts.

The next response from industry was a massive wave of mergers and acquisitions, in which many companies simply disappeared into huge new conglomerates. The attempt to deal with the intensity of competition by forming trade associations to police behaviour had failed. Instead, competition was simply reduced by firms eliminating rivals by merging them into a single organization.

In part, this dramatic reduction in the number of major players in each market was triggered by another piece of legislation, the general incorporation laws passed by the state of New Jersey in 1889. General incorporation laws. The very phrase triggers a colossal thud of boredom in the minds of most people. But astute business people realized that they provided a way round the fundamental problem facing trade associations, namely, the incentive for any individual member to renege in secret on the deal and to cut its prices. The precise details of the laws need not concern us, but they gave ingenious companies the means with which to enforce legally the intra-firm agreements on prices and output.

The main pressure towards the elimination of rivals by merger and acquisition was, however, the continued interest of legislators in restricting any form of agreement or collusion between individual firms. This was a major theme in American domestic politics in the decades leading up to the First World War.

This merger movement, by far the largest of its kind up to that time, lasted from around 1895 to 1905. During that time, it is estimated that 3,012 firms, most of them of substantial size, disappeared because of mergers. The value of the consolidated firms which emerged as a result totalled almost $7 billion, or well over $1,000 billion at today's prices.

Consider the corporate world on the eve of the First World War. Firms had sprung up which were of wholly unprecedented size. For the first time ever, individual companies both could and did operate, not just on a continental, but on a global scale. This new and dramatic phenomenon attracted the attention of a shy English academic who was probably the most innovative and influential economist in the world at the time.

Alfred Marshall spent most of his life cloistered in the rooms of St John's College, Cambridge. Trained originally as a mathematician – indeed he was placed second in the whole university in his final exams – he switched his attention instead to economics. His *Principles of Economics*, revised through many editions, became what was probably *the* dominant book in the rising discipline of economics in the early decades of the twentieth century. He was instrumental in persuading a new Cambridge intellectual star, John Maynard Keynes, to focus on economics.

Despite his apparent other-worldliness, Marshall was an acute observer of contemporary economic and business life. His *Principles* are littered with insights, some elaborated at length, others merely mentioned in passing, which remain interesting and thought-provoking even today. An important question that Marshall considered was: how long will these new giant firms live? His interest in this matter is instructive to many economists today. Rather than busy himself with abstruse theoretical models whose assumptions lack empirical validation, Marshall, like all the great economists before him, was concerned to try to answer questions which illuminated the main economic issues of the day. He was more than capable of formulating theoretical models, as his early prowess as a mathematician testifies, but he regarded them as tools to help understand how the world works. The emergence of the giant corporation was a new phenomenon, which required analysis and understanding.

Marshall revised and altered his opinion on the survival of mega-corporations in successive editions of his *Principles*. In the first edition, published in 1890, he argued that, like trees in the forest, there would be large and small firms but 'sooner or later age tells on them all'. But by the sixth edition in 1910, his view had changed. It is during these two decades, remember, that the truly massive companies had emerged. Perhaps a little star-struck by their sheer size, Marshall then opined that 'vast joint stock companies ... often stagnate, but do not readily die'. He believed that these new firms were qualitatively different from the nineteenth-century firms on which he based his previous generalization.

With the benefit of hindsight, we can see that, in this particular case, even Homer nodded. Marshall's original insight of 1890 has proved to be much nearer the truth than his revised opinion of 1910. Even giant firms fail.

Much of our information on the experiences of these companies in the twentieth century comes from two impressive and detailed studies by the American sociologist Neil Fligstein and the British economic historian Leslie Hannah. Fligstein's book, *The Transformation of Corporate Control*, provides evidence on the lives of the 100 largest companies in America at the end of each decade from 1919 to 1979. Hannah's article* traces the survival of the world's 100 largest industrial companies in 1912 through to 1995.

In recent years, the collapse of corporate giants such as Enron or WorldCom has caught the headlines in a spectacular way, with investors avidly searching for scapegoats. The level of scepticism about the pronouncements of companies has risen. In many ways, this is a healthy sign. As I write these words, for example, it emerges that the oil company Shell has overstated its proven oil reserves by 25 per cent. A byword for boring conservatism, the company grew complacent and arrogant. The chairman, Sir Philip Watts, has already been compelled to resign, and the market value of the company has fallen by over $15 billion in less than two months. No fewer than four chairmen of FTSE 100 companies, all previously senior executives of Shell, are amongst those being pursued in the US courts for damages arising from the mis-statement of reserves.

Despite this recent bout of scepticism, most commentaries about the business world continue to eulogize success. But as Hannah

* L. Hannah (1999), 'Marshall's "Trees" and the Global "Forest": Were "Giant Redwoods" Different?' in N. R. Lamoreaux, D. M. G. Raff and P. Temin (Eds), *Learning by Doing in Markets, Firms and Countries*, National Bureau of Economic Research.

notes laconically, 'The tendency to overemphasize successes, and to rationalize them *ex post* is chronically endemic amongst business historians and management consultants.' The latter group are particularly prone to the temptation of claiming to have found the unique formula for business success. Books proliferate, and occasionally sell in very large numbers, which claim to have found *the* rule, or small set of rules, which will guarantee business success. But business is far too complicated, far too difficult an activity to distil into a few simple commands, be it the 'set price equal to marginal cost' of economic theory, or some of the more exotic exhortations of the business gurus. It is failure rather than success which is the distinguishing feature of corporate life.

We see the survivors, and their triumphs are lionized. But the failures remain virtually forgotten. Hannah's list of the world's largest industrial companies in 1912 contains familiar names: Procter and Gamble, Siemens, General Electric and, yes, Royal Dutch Shell, to give the currently afflicted company its Sunday-best name. But there, too, is Briansk Rail and Engineering, not long for this world after the Bolshevik seizure of power in Russia in 1917. And Hohenlohe Iron and Steel of Germany along with Central Leather and Cudhay Packing in the US. Gone. Gone and forgotten.

Fligstein's evidence is less detailed than Hannah's for our immediate purposes, though it contains much interesting material. His data set does not include evidence on whether a firm failed completely and ceased to exist as an independent entity. Rather, it focuses on whether or not a company was in the list of the largest 100 US firms at the end of each decade from 1919 to 1979. Only thirty-three out of the top 100 in 1919 remained in the list in 1979, and since then the attrition amongst the survivors has continued.

Fligstein notes that no fewer than 216 companies in total made it into the US top 100 over the sixty-year period. Some, such as Bethlehem Steel, WF Woolworth, Chrysler and Goodyear Tire and Rubber were in the list for the entire period. Others enjoyed their fifteen minutes of fame in a single appearance, such as Atlantic

Gulf and West Indies Shipping Line in 1919, Lehigh Valley Coal in 1929, Climax Molybdenum in 1939, Allied Stores in 1949, Kaiser Steel in 1959, International Utilities in 1969 and, anticipating the future, Rockwell International in 1979. International Business Machines (IBM) makes its first appearance in 1939, but otherwise computing firms such as Microsoft are absent, simply because for the most part they barely existed at the last date on Fligstein's list, 1979.

On average, over the individual decades from 1919–29 to 1969–79, seventy-eight out of the top 100 at the start of any decade were still there at the beginning of the next. But no fewer than twenty-two out of 100 were not. These are, or rather in most cases were, the giants of American capitalism. Operating on a massive scale, and possessed of enormous resources, almost one in every four were unable to remain in the top 100 for more than a decade.

Hannah's evidence takes us to a more recent date, 1995, and shows not only when firms merely dropped out of the world's top 100 but also when they ceased to exist as independent concerns.

The companies in the world's top 100 in 1912 represented the cream of capitalism. These were the survivors of a brutal era of competition and had successfully survived the massive wave of mergers around the turn of the century. As Hannah points out, 'They were, on the whole, firms that contemporary stock-market analysts considered attractive and safe because of their consistently reliable record of generous but sustainable dividends. A population of the largest firms of ten years earlier would almost certainly show earlier exits and faster rates of decline than this population.' In short, these were the blue-chip companies of their time. The value of the smallest, in stock-exchange prices of 2004, was $5 billion, and of the largest $160 billion. Yet within ten years, ten of them had disappeared as independent concerns. Fitness, in the form of huge assets and years of successful operation, proved no guarantee, not merely of continued success, but of their very survival. Their experiences over the period 1912–95 are summarized in Table 1.1.

TABLE 1.1 Performance of the world's largest 100 industrial companies in 1912 over the period 1912–95 (*Source*: Hannah, 1999)

Bankrupt	29
Disappeared	48
Survived	52
Remained in top 100 in 1995	19

The period 1912 to 1995 is not much longer than the average life expectancy of a human being. Of course, life expectancy was shorter in 1912 than it is now, but many members of the middle class born in 1912, with their greatly superior diet, better housing and less arduous working conditions, could quite reasonably have expected to live until 1995. So, within the span of a human life-time, each of the twenty-nine companies that became bankrupt squandered many billions of pounds in shareholders' funds. Money which existed as the value of the shares in these companies vanished to nothing. How hard is it to spend so much money? For an individual, there are only so many apartments in Manhattan and Mayfair you can buy, only so many great works of art, only so many private jets and yachts. Yet no fewer than twenty-nine out of the world's top 100 companies in 1912 succeeded in making stupendous sums of money disappear.

In total, forty-eight out of the top 100 disappeared as inde-pendent entities, and only twenty-eight were larger in 1995 than they were in 1912. A small number of the companies, such as Procter and Gamble and BP, were very much larger, expanding shareholder value by a factor of at least seven. But these were the exception rather than the rule. Disappearance or decline was almost three times as likely as growth. Because of my own back-ground, I am often asked by would-be entrepreneurs seeking escape from life within huge corporate structures: 'How do I build a small firm for myself?' The answer seems obvious: buy a very large one and just wait.

We do know that the average lifespan of small firms is shorter than very large ones, but this is accounted for almost entirely by relatively high death rates in the very early years of existence. The offer, launched with such excitement and anticipation, turns out to be not quite right for the intended market. And, unlike a large company, the very small firm is not diversified. The failure of its main product means the end of the firm itself. At a more elementary level, the owners of the firm may simply get their cash flow wrong and not be able to meet in time the demands of ruthless predators such as the tax authorities. But, after the first few years of existence, the difference between large and small firms' survival potential narrows dramatically. Their prospects of surviving the next year become more or less the same. And, eventually, age claims them. Most firms fail.

We can see this from the very dawn of the modern age. The spread of printing in Europe in the late fifteenth and early sixteenth centuries was one of the greatest technological leaps ever made, with an impact far more dramatic and pervasive than that of the internet. Information and knowledge could now be disseminated widely for the first time in human history, freed at last from the constraints of the need to transcribe scripts by hand in order to make a copy. Intellectual ferment and fervour was at its height in the Italy of the Renaissance, and nowhere more so than in Venice, centre of a vast network of international trade. In 1469, twelve companies were engaged there in the new activity of printing, but by 1472 nine of them had failed.

Moving to more modern times, the domination of the world car market by a relatively small number of very large firms seems an immutable fact of life, but between 1900 and 1920 there were almost 2,000 firms involved in automobile production in the US. Over 99 per cent disappeared. Likewise, Hollywood now bestrides the world of film-making, yet in the 1900s the European film industry exported throughout the world, at times supplying half the US market. By 1920, however, European films had virtually disappeared from America and had become marginal in Europe.

Both the evidence from the first giant wave of mergers around 1900 and the experiences during the twentieth century of the great success stories which emerged from it are entirely typical of the patterns of behaviour traced by firms. They innovate; they respond creatively to changes in their external environment; they strive not merely to survive but to succeed. Some do, but for the most part they fail.

This chapter has looked at how hard it is to build successful long-term businesses. Firms have to operate in the uncertainties of the market. In the next chapter, we examine what economics has to say about markets, firms and how they behave. We will see that economic theory neglects almost completely the widespread existence of corporate failure. It has a great deal to say, not about what firms actually do, but what they ought to do in order to succeed. And we will see that the most widely quoted recipe for success taught in the economics textbooks is, in many real-world situations, a prescription for failure.

2   A Formula for Failure

It is easy to be critical about economics. P. J. O'Rourke, in his entertaining book *Eat the Rich*, has a stimulating definition of the content of the subject: 'One thing that economists do know is that the study of economics is divided into two fields, "microeconomics" and "macroeconomics". Micro is the study of individual behaviour, and macro is the study of how economies behave as a whole. That is, microeconomics concerns things that economists are specifically wrong about, while macroeconomics concerns things economists are wrong about generally.'

As far as most economics textbooks are concerned, O'Rourke's definition seems perfectly reasonable. Failure is all around us, a pervasive feature of everyday life. Government policies fail, firms fail, whole economies fail and remain enmeshed in poverty. Failure is both general and specific. As Lucy Kellaway, an award-winning columnist in the *Financial Times*, recently wrote, albeit slightly tongue-in-cheek: '83 per cent of Chief Executive Officers fail'.

The existence of failure on this scale is simply not recognized in economics. Instead, in much of economic theory, getting the right strategy, the right policy, is straightforward. It is simply a matter of following the appropriate formula laid down in the textbook. The formula itself may be expressed in varying degrees of mathematical complexity, depending upon the level of student at which the text is aimed, but it will, in essence, remain the same.

Economics as an academic discipline does have a very valuable strength: it trains people to think analytically. And, as we will see during the course of the book, the subject has not stood still over the past thirty years. Important progress has been made in understanding certain aspects of how the world behaves. But, in essence, economic theory remains rooted in a vision of the world which

17

was derived from the physical sciences of the nineteenth century. The achievements of the latter are plain for all to see. The analytical techniques and mathematical tools used by nineteenth-century scientists enable us to understand a great deal of the world around us.

These have been much less successful, however, when applied to human social and economic systems. The fundamental reason for this is that this approach regards equilibrium – a static, changeless state of the world – as the natural order of things. The whole panoply of differential calculus, the branch of mathematics that is by far the most widely used in economics, is focused on finding equilibrium solutions, solutions in which the system is at rest, static, in which continuity and lack of change are its hallmarks.

This is simply not the case either with society or with the economy. As we saw in the opening chapter, one of the most obvious features of firms in the real world is that they fail. To remind ourselves of just one piece of evidence, over 10 per cent of all companies in the US, the largest and most successful economy in the history of the world, fail every single year. An analytical approach that at heart involves the concept of equilibrium cannot really cope with this most dramatic feature of change, namely failure. One year a company exists, the next it has disappeared. Economic theory fails to reflect this fundamental point. There is little or no mention of failure. Instead, the emphasis is on what firms need to do not just to succeed but to take the best possible decision from the entire range of options which is available to them.

In the textbook world, running a business is easy. Almost all the problems which exercise management are swept away. Rather, they are never mentioned at all. Despite this, economists have an irritating habit of claiming the discipline is the first to explain anything worth knowing about how the business world works.

A single example will suffice to illustrate the point. A successful textbook, by highly respected authors and specifically directed at the interface between economics and business, contains an amazing opening remark. 'The economic analysis of firms,' it is declaimed, 'provides a simple decision rule that managers and entrepreneurs will find useful.' What remarkable insight is about to be vouchsafed to us? It is none other than the truly original concept that 'Any action that adds more to revenue than it adds to cost should be undertaken,' and, what is more, 'Actions that add more to costs than to revenues should not.' The text goes on, 'In the economic theory of the firm this rule is heavily disguised.'

By this I think the authors mean that you have to spend years acquiring at least first-year-degree-level maths before you can begin to understand it.* Alternatively, we may regard this rule as an illustration of the technique outlined by the British satirical writer Stephen Potter about how to gain the upper hand in a conversation about business. In his excellent book *One-Upmanship*, he describes his 'Economics B' technique as the 'Approach of Utter Obviousness'.

The business-school gurus, it should be said as an aside, are by no means immune to the temptation to reduce highly complex problems to a set of easy formulas. Leading figures like Tom Peters offer beguiling sets of simple rules. Business-school thinking has fallen into the trap of standard economics: everyone can use the same rules and prosper. But most of the firms in Tom Peters' *In Search of Excellence* later failed to maintain their excellence, to say the least.

An important defect – a failure, I am almost tempted to say – of the equilibrium approach in economics is that it tells us nothing about the timescale of the process of change from one equilibrium to another, even in the textbook world in which equilibria exist by definition. In reality, of course, as we have noted above, they might not exist at all.

But in the comforting world of economic theory, we can think, for example, of a single market in which the price of a product is set so that supply and demand balance. The amount which firms

---

* To be fair, this particular book does get considerably better as it goes along and is one of the very few to make an effort to relate to real-world business problems. It is *Economics for Business and Management* by Alec Chrystal and Richard Lipsey (Oxford University Press, 1997).

are induced to supply to this market at this price is equal to the amount which consumers are willing to buy. This is one of the most basic features of any economics textbook, even at the most elementary level. As it happens, as we shall see below and again in Chapter 5, even this apparently straightforward concept gives rise to some difficult and awkward questions for conventional economics.

But the relevant point here is that the emphasis in economics is on a description of such equilibrium situations, in which supply and demand balance exactly. If the system receives a shock – a tax being imposed, say, on the product in question, so that its price goes up – economic theory describes the new levels at which supply and demand will once more be equal. Both producers and consumers adjust their decisions in the light of the new price. We can then compare the two equilibria, both before and after the tax was imposed. The theory tells us nothing at all about how long the system will take to move from the old equilibrium to the new one, following the introduction of the tax. And it says nothing at all about the path the system follows between these two equilibria. Is most of the adjustment made rapidly, for example, and the last few steps taken more slowly, or is the path one of entirely smooth progress? Nobel laureate Vernon Smith of George Mason University made the point clearly in his Nobel lecture of 2002.

In the context of a very basic market, such as cornflakes or bread, for example, these questions might not really matter in practice. But once we start thinking about more complicated problems, the process by which the system moves from one equilibrium to the next, and the time this takes, become very important issues indeed.

For example, economics has a theory of growth, growth being the single feature that most distinguishes the market-oriented economies of capitalism from all other previously and actually existing societies. No other system has ever generated such steady growth over such a long period of time. For example, the increase in output (after allowing for inflation) in the western world between 1950 and 1973 has been estimated to have been twice as much as in the period of almost two millennia between o and 1800 AD. So growth is a pretty important topic, to say the least.

As long ago as 1969, the then young Cambridge academic Tony Atkinson, now recently retired as Warden of Nuffield College, Oxford, investigated the timescales of various models in economic theory. In particular, he looked at equilibrium solutions in the standard theory of growth and calculated how long it takes for a system to move to a new equilibrium once a change is introduced. He found that the answer was typically over a hundred years. This is not a misprint. One hundred years. In other words, even within the safe confines of economic theory, a system which is in equilibrium will take more than a century to move to a new one once a shock is administered to it. So the system spends a long time in *dis*equilibrium, when things alter and no longer run completely smoothly.

Economic and social systems are essentially dynamic and not static. Even if we were to retain a faith – for that is what it is – in some Platonic idea of equilibrium towards which the economy is moving, most of the time we will experience and observe behaviour not in but out of equilibrium. Firms fail, policies alter, behaviour changes. A theory which is based on describing what the economy is like once it reaches an equilibrium, a static solution, will of necessity give us at best only a partial understanding of the world.

In practice, firms fail for a whole variety of reasons, some of which might be common to a number of firms, such as an economic recession causing a drop in sales, or which might be specific to an individual company, such as losing sales to a more effective competitor. But the proximate reason for failure, the thing that happens just before they fail, is that firms run out of money. They do not have enough revenue to cover costs and are unable to beg, borrow or steal sufficient funds to fill the gap.

Firms raise revenue by selling their products, whether goods or services; a statement of, in Stephen Potter's phrase, the utterly obvious. Equally obvious is that revenue depends upon how much of the product is sold, and what price it is sold at.

Choosing the price of a product is a challenging and difficult decision for companies. If a firm makes a big enough mistake on this, or even persists with perhaps relatively minor mistakes for a sufficiently long period, it will fail. Many factors need to be considered. The same product need not have the same price at the same time: there might, for example, be discounts for bulk buying or special terms for either loyal or new customers. And who, for example, are the main competitors, and how might they react to a change in the price of your product? Are there any potential competitors likely to enter your market with a rival offer or, worse, be developing one which is better than yours? How can pricing strategy be used to deal with the more shadowy but often dangerous threat of potential, as opposed to actually existing, competitors?

In the world of textbooks, the formula that most economics students learn reduces these and other difficulties to a single phrase: 'set price equal to marginal cost'. The word 'price' is used in its everyday sense, but the phrase 'marginal cost' is a piece of jargon. 'Marginal' is a word that permeates the whole of standard economic theory. It has an entirely specialized meaning in this context. A workaday translation is 'additional'. So, stripped to its essentials, the phrase 'marginal cost' means the cost of producing an additional unit of output.

We might usefully recall here the piece of advice given by economists to decision-makers in companies: 'Any action that adds more to revenue than it adds to cost should be undertaken.' If we are able to sell an additional unit of our product, the additional revenue we obtain from it will be the price at which it is sold. 'Marginal cost' is the cost of producing an additional unit of the product. So if we are able to sell another unit when the extra revenue is greater than the extra costs we incur in producing it, we should do so. And we should go on doing so until we reach a point where the additional revenue, the price, is equal to the additional cost of doing so. This is what the formula 'set price equal to marginal cost' means.

By following this formula, according to the textbook model, firms will automatically succeed in making the maximum amount of profit available to them. The formula can be expressed at different levels of sophistication. It can be demonstrated graphically. It can be shown using differential calculus. It might even, for graduate students, involve a knowledge of Brouwer's extension of Kakutani's fixed-point theorem in the proof of the existence of general equilibrium. The formula is by no means purely of academic interest. It is the principle that underlies a great deal of the activity of regulatory authorities around the world, and is the standard against which potentially anti-competitive behaviour is often judged.

But, regardless of the mathematical language in which it is couched, it is not our concern here to prove that the formula does indeed lead to a firm making not just a profit but the maximum amount possible. Any non-economist reader who wishes to understand this can readily consult any one of a vast range of textbooks. Rather, our concern is to discuss the formidable difficulties of following this rule in practice. Many factors which can and do lead to failure in reality are entirely missing from this way of thinking.

The single most important difficulty with this rule of how to set price is the amount of information, of knowledge which the firm is assumed to have. The company is presumed to understand exactly how many sales it will be able to make at each particular level of price. In the jargon of economics, the firm knows its demand schedule or curve. From another aspect, the firm is required to know its cost curve, or how the costs of production vary with the level of production.

These may seem at first sight to be innocuous assumptions, but the practical difficulties of obtaining this information with certainty are enormous. Consider, first of all, the demand curve, the information which in principle will tell the firm how much of a particular product people will buy at any given price.

Two enormous, related industries, advertising and consumer market research, have developed which attempt both to understand the complicated motivations of consumers and to try to shape them in the interests of their client, the producer. If the

23

process of running a business were easy, if competitors were completely obvious, if all the factors which attract consumers to a product were known, if the reaction of sales to price were understood exactly, advertising and market research would scarcely be necessary. The very existence of these industries on such a large scale is testimony to the pervasive nature of uncertainty in business, even as regards such a basic concept as setting the price of the product. And when uncertainty exists on such a scale, the potential for getting things wrong, for failing, exists alongside it.

It is worth mentioning at this point a distinction that is made in much technical work between the concepts of risk and uncertainty. In casual, everyday English, the terms might be used more or less synonymously. But in their specialized usage, particular within economics, there is a difference to be drawn. Risk refers to situations in which the outcome cannot be known with certainty, but the probability of any given outcome *is* understood perfectly. A simple example would be a toss of a fair coin. There is a fifty–fifty chance of it being either heads or tails. If we are gambling on the next toss being heads, there is a risk that we will lose our money if it turns out to be tails. But we know precisely what the chances are. Uncertainty, in its strict sense, refers to situations in which the probability of the various outcomes is itself unknown. So, for example, we might place a bet on the Earth being visited by creatures from outer space in the next ten years. Here, no one knows for certain what is the true probability of this happening. Almost everyone would agree the chances are very small, but are they 1 in 100 million, 1 in 100 billion, or are they literally zero?

Modern economics deals extensively with the concept of risk. Risk involves quantifiable probabilities. This makes it amenable to mathematical manipulation, a practice dear to the hearts of economists. But the great economists of the past wrote much more about situations involving genuine uncertainty, regardless of their own ideological persuasion. Keynes, for example, believed in more extensive government involvement in the economy. Equally, however, he wrote a great deal about situations which were uncertain rather than merely risky. Frank Knight, who founded the free-market school of economics at the University of Chicago and who Milton Friedman described as 'one of the most original and influential social scientists of the twentieth century', also thought instinctively about uncertainty rather than risk. His doctoral dissertation in 1921 was indeed the seminal piece of work to make this distinction. Interestingly, in the context of this book, Knight famously responded to his own rhetorical question 'How far is life rational?' with the blunt answer 'Not very far.'

In practice, of course, the two concepts almost always blur into one another. In practice we rarely face situations which are as clear-cut as the outcome of the toss of a coin. Equally, it is unusual for us to have no idea whatsoever about the likely distribution of possible outcomes. The precise mix will depend upon circumstances.

Returning to the problem of companies grappling to understand their customers, both actual and potential, an immense amount of effort and money is expended by large firms in trying to understand the shape of the demand curve. Focus groups, sophisticated consumer surveys, complicated statistical modelling of data are all used in an attempt to reduce a fundamental uncertainty confronting any business, namely, what happens if I change the price of my product? But despite the striving for knowledge, uncertainty persists.

Decision-makers in companies are particularly uncertain about how their competitors might react to any changes in price that they might make. Reducing prices to boost sales might make sense, but not if your key competitors follow suit and the end result is that everyone has to charge a lower price and hence make less of a profit margin on sales. Judging how competitors might react takes us to the concept of game theory, an esoteric idea made famous to the general public in the film *A Beautiful Mind*. The film tells the story of the eccentric American academic John Nash, who invented one of the fundamental propositions of game theory. But game theory is such an important idea, both in its own right and in terms of what it can and cannot tell us about failure, that it merits a prolonged discussion of its own in Chapter 5.

The uncertainty which firms face in understanding the nature of the demand schedule for their product is pervasive. The day I was writing the first draft of this chapter in 2003, in the agreeable winter climate of New Mexico, the newspapers carried two such stories. At what appears to be a mundane level, but all the more important for being so normal, were the stories in the American press about the Japanese vehicle manufacturers' strategy for capturing part of the lucrative US market for pick-up trucks. This is a highly profitable industry, with US sales of just over two million a year bringing in almost $20 billion profit. Until recently, the large American manufacturers faced no challenge in this market from the Japanese. Now, a dangerous new competitor was starting to roam across the pastures.

In early 2003, Ford launched their redesigned F-150 pick-up, the most popular vehicle in America in this market for over twenty years. The very next day, Nissan unveiled their first full-size pick-up, the Titan. The Nissan Motors CEO was quoted as saying, 'We have done our homework and know what the big pick-up buyer wants.' But the General Motors' vice-president in charge of pick-up trucks thought otherwise. Gary White conceded that, in the 1970s, Japanese cars made inroads into the US market because 'we had the wrong products and we had poor quality'. He insisted that this time it was different, a view echoed by the Chrysler Group market-research director Dave Bostwick: 'We paid too little attention and let these guys in the door in the past, but this time I think everybody's ready for them.' But GM overall product chief Bob Lutz was prepared to concede that a danger existed: 'The risk for us is if consumers prefer Nissan styling and their power trains.'

So here were some of the largest industrial companies in the world, each with enormous experience of the market, each spending vast amounts on trying to discover what their individual demand schedules look like or, in plain English, how many pick-ups they would sell in 2003. Nissan and the other Japanese firms thought they would do well and gain significant market penetration. GM and the American firms thought not. Both of them could not be right. One or the other of these competing groups, or maybe both, had an imperfect knowledge of the demand schedule which they faced. In advance, before the event, no one knew for sure who was right and who was wrong.

It is worth repeating the quote given by Bob Lutz at GM: 'The risk for us is if consumers prefer Nissan styling and their power trains.' How might we expand and deconstruct this statement, possibly prepared at considerable expense through the public-relations advisers? The following seems a fair attempt: '*If* it turns out to be the case that consumers like what the Japanese have to offer more than the stuff they are used to buying, then they will buy the Japanese product. But if they don't, then they won't. And I don't really know whether they will or not. I can only say that there is a risk of it happening.' My intention is not to mock Mr Lutz's statement but to pay him due honour for recognizing so clearly and so openly the fundamental uncertainty about the demand schedule that companies face. And if GM, with its vast resources, does not really know, if GM cannot tear down completely the veil of uncertainty that shrouds the future, who can?

Certainly not the airline industry, which featured in the second of the stories in the newspapers. The world's airlines lost $13 billion in 2002, an improvement upon the $18 billion loss in the aftermath of September 11. And despite enormous increases in the number of air travellers in the second half of the twentieth century, the combined profit made by airlines over the period 1955–2000 was less than the losses they incurred in just the two years 2001 and 2002. These losses are not surprising, bearing in mind the attack on the World Trade Center. But this was, admittedly in dramatic form, an example of the sudden changes in the external environment which can buffet even a well-run company.

An example from the European market emphasizes the point. In the last few years, the cartel of national air carriers that dominated the market for so long has been challenged aggressively by low-cost, no-frills airlines. The major airlines were eventually forced to respond, both in terms of their own pricing strategies and in terms of setting up their own, dedicated low-cost operations. The Dutch company KLM set up Buzz, which incurred multi-million pound losses in each of its three years of existence before it was sold in January 2003 to another low-cost airline, Ryanair of Ireland. The price for the company, after deducting its cash balances, was a mere £4 million. As Gert Zonnenveld, airlines analyst at WestLB Panmure, said, 'It sounds very easy to say, "Let's start up a low-cost airline and make lots of money." But in reality it's not easy, and it takes quite a few years – if you are lucky.'

There have already been numerous firms that have failed in this sector. Ciao Fly began operations in the summer of 2002. Described as 'a Swiss company run by a German, operating an Italian airline to Britain', it lasted just six weeks. Swedish entrepreneurs launched Goodjet in April 2002. Visitors to its website are now directed to the site of a firm of accountants, where a sad little note reads, 'Goodjet was declared bankrupt on 17 January 2003 by the Gothenburg district court.' Even the doyen, if the term might be used in this context, of European low-cost flying, Easyjet, is currently making investors nervous as its seat occupancy falls sharply.

It seems clear, then, that business is not as easy as the economics textbooks pretend it is. The whole process of trying to penetrate the uncertainty that surrounds identifying all your key competitors, quantifying the impact of pricing strategies both by you and by your competitors on how much you will sell, understanding the effect of advertising and promotion, solving the problem of distribution to actually get your products in the shops for customers to buy – this entire process, enmeshed in uncertainty, is assumed away in the textbooks. You *know* what your demand schedule looks like. In other words, you know how much of your product will be bought at whatever particular price you decide to charge.

Suspending for a moment the disbelief that all this information has appeared as if by magic, if we know the demand curve, we know how much extra revenue we can get if we are able to sell one unit of output in addition to our current level of output. And we know this for all possible current levels of output.

The formula 'set price equal to marginal cost' means, we recall, that we should produce exactly that level of output at which the additional revenue obtained from selling an extra unit of output is equal to the additional cost of producing it. So we also need to discover how our costs of production vary with the level of our output.

Unlike the question of finding out about how the demand for a company's products might vary as things like price are changed, this task does not seem so unreasonable. The process of production is internal to the firm. The company is not faced to the same extent by the vastness of the external world and the associated uncertainty as it is when it thinks about how its products might be sold.

Even so, the discovery of how costs vary over different ranges of output might not be completely straightforward. For example, the British government recently brought in a scheme to assist in meeting pollution reduction targets under the Kyoto protocol, specifically for the reduction of carbon dioxide production. Companies can participate in what is known as the emissions trading scheme. Under this, the firm must undertake a binding agreement to reduce its use of energy – and hence its emissions of carbon dioxide – by a specified amount by a certain date. In return for this commitment, it receives allowances to produce emissions at the agreed level. The real innovation, however, was that these permits can be traded. If a company finds that it proves easier to meet its target than it originally thought, it will be able to sell some of its permits. Its emissions will be below the level which it contracted to 'produce'. Equally, if a firm miscalculates in the opposite direction and discovers that it is much harder and more expensive to cut its emissions, it can buy permits which allow it to produce emissions at a higher level than its original target.

Now, if firms knew exactly how the costs they incur in reducing the energy requirement of their production processes varied, from their point of view these allowances would not be required. No firm would make a mistake. But the scheme has proved very popular in practice, with very large companies being active participants. Like Sherlock Holmes's dog that did not bark, this tells us a great deal. It tells us that the companies which use large amounts of energy and which participate in the trading scheme realize that they *can* make mistakes about their knowledge of how their own costs evolve.

Just as we imagined earlier that a firm knows its demand curve, suppose as well that it understands its cost curve. In other words, it knows what its costs of production are at all possible levels of output. The management can then work out – or get some accountants to do it for them, thereby adding even more to the costs – how the costs per unit of output change as the scale of production varies. It is this change that is key to the concept of 'marginal cost'. Marginal cost means the cost of producing an additional unit of output. So, in principle, we can calculate, starting from every single possible level of production, the cost of producing the next unit of output.

Perhaps we are finally there. Perhaps we can now put into practice the formula 'set price equal to marginal cost'. We know the cost of producing an extra unit of output over and above any particular level of output. We know the revenue we can get from selling an extra unit of output over and above any particular level of sales. So we find the point at which the two are equal and – abracadabra! – we know not only the exact amount to produce, but we know that we will make the maximum possible amount of profit, given our demand and cost curves.

But the practical difficulties are not yet exhausted. In many situations, the higher the price which is charged for a product or service, the less of it will be bought. So, in the jargon, the demand curve slopes downwards. If we draw a chart with the level of sales on the left-hand axis and the price on the bottom axis, the demand curve will slope down from left to right across the chart. At low levels of price, demand will be high and at high levels, it will be low.

We can immediately think of exceptions to this. In financial markets, for example, it is often the case that, when the price of a share rises, traders buy more of it and not less, since they imagine that the price will continue to rise and so they can sell it later at a profit. This motivation was particularly strong during the stock-market boom of the late 1990s. In consumer markets where high fashion plays a key role, a higher price might serve to attract rather than deter those individuals who enjoy conspicuous consumption. At the other end of the scale, a second-hand car that seemed to be priced too cheaply might very well not sell at all, on the grounds that if it is so cheap there must be something wrong with it. But, in general, we seem on reasonably safe ground if we assume that the demand curve slopes down; that the lower the price, the more will be sold.

We are almost, but not quite, there. We need to know a little bit about the shape of the marginal cost curve, how marginal cost varies at different levels of output. The standard assumption made in teaching generations of students of Economics 101, the introductory course to university economics, is that as the level of output increases, marginal cost first of all falls but eventually rises.

At first sight, this might not seem unreasonable. We might imagine a car plant being built. The cost of producing just one car is enormous – all the costs of the plant are allocated to this single car. When two are produced, the costs of producing the second car are dramatically less. So the marginal cost of making the second car is much less than that of the first. And as more are produced, the marginal cost of each continues to fall. But at some point, the capacity of the plant will be reached and producing an additional car will need a whole new plant – costs per unit of output will once again rise. Even before then, there may well be bottlenecks in capacity. The workers may require overtime payments at premium rates, for example, or a particular segment of the production line might require upgrading. Such factors will lead to an increase in the cost of making that extra car, even before a whole new plant is needed.

Focus now on the part of the marginal-cost curve where it starts to slope upwards. At levels of output lower than this, price must be higher than marginal cost, otherwise we make no profit. Indeed, if marginal cost is higher than price over the entire downward sloping part of its range, there is no point in producing at all, because each additional unit of output makes a loss. So when marginal cost starts to rise, at some level of output, having been less than price, it becomes equal to price. This is the magic spot, where price equals marginal cost and by definition no profit is made on that particular unit of output.

So there we are! We can all go off and happily maximize our profits like all good capitalists. Or can we? The problem is that in substantial sections of the economy it is not at all clear that the marginal-cost curve ever gets to the point where it starts to slope upwards rather than downwards.

An obvious contemporary example is when a company offers subscription services over the World Wide Web. Once the process of debiting credit-card payments has been automated, the material set up in an appropriate form and the initial teething problems of running the system overcome, the marginal cost – the cost of selling to an additional subscriber – is virtually zero. The marginal-cost curve never slopes upwards. On the contrary, it approaches zero. A company which charged marginal costs in these circumstances would soon go bankrupt, because it could never recoup the costs involved in setting up the service in the first place.

More generally, firms strive to avoid a situation in which the marginal-cost curve starts to slope upwards. The managers may not see it in precisely the jargon of economics, but a great deal of their time and effort goes into trying to provide their product more efficiently: their companies carry out research and development, they invest in new equipment, they look for cheaper and better sources of the supplies they need to produce their product, they evaluate their promotional and marketing material, and so on.

An important and mysterious effect that seems to occur in many industries is that the costs of producing a given level of output fall, the greater the cumulative level of production has been. In other words, simply by producing more of a product over time, the process of production becomes more efficient. As long ago as 1936, an article innocuously titled 'Factors Affecting the Cost of Airplanes' provided empirical evidence that the direct labour cost of manufacturing an airframe fell by 20 per cent with every doubling of cumulative output. Many further studies have found a similar qualitative effect in a wide range of industries, although the exact gains in efficiency differ.

All these activities mean that the marginal-cost curve is not the fixed, immutable line of economic theory. It is the outcome of a dynamic process, and it changes and evolves over time. The quest to produce a better product more efficiently is the driving force behind all successful businesses.

Interestingly, this was a theme emphasized again and again by some great economists of the 1920s and 1930s. Allyn Young, for example, who held professorships first at Stanford, then Cornell and finally Harvard before dying tragically early aged just fifty-three, and who supervised Frank Knight, whom we met briefly earlier in this chapter. Young wrote a classic article in 1928 in what was then the world's leading academic economics journal showing how falling marginal-cost curves undermined the traditional economic concept of equilibrium. Joseph Schumpeter, another Harvard professor, whom we will meet again in the final chapter, emphasized that the key features of actually existing western economies were change and discontinuity, not equilibrium.

Piero Sraffa, a Fellow of Trinity College, Cambridge, published an article in 1926 arguing that in very few industries indeed did marginal-cost curves ever slope upwards. He ridiculed the importance that conventional economics attached to this concept, writing, 'Business men, who regard themselves as being subject to competitive conditions, would consider absurd the assertion that the limit to their production is to be found in the internal conditions of production in their firm, which do not permit the production of a greater quantity without an increase in cost. The chief obstacle against which they have to contend when they want gradually to increase their production does not lie in the cost of production – which, indeed, generally favours them in that direction – but in the difficulty of selling the larger quantity of goods without reducing the price, or without having to face increased marketing expenses.'

Sraffa, like many Cambridge dons of his generation, held a lifelong political attachment to Stalinism, but he did not allow this to influence his own conduct. For example, in 1945, after the atomic bombs had been dropped on Hiroshima and Nagasaki, he invested all his money in Japanese government bonds, whose price then was obviously very low indeed. He judged that capitalist Japan would recover, despite the apparently devastating blow it had just suffered. And indeed it did, surpassing its pre-war levels of output by 1953 and going on with thirty more years of stupendous growth. Sraffa made a fortune.

So how *do* firms actually behave when it comes to deciding on the price of a product? Immense efforts have been made in economics to develop and make more precise and rigorous the whole corpus of theory associated with the phrase 'price equals marginal cost'. The theory is distilled and passed down to generations of students at all levels of the discipline. Yet the empirical evidence suggests that firms rely in practice much more on simple rules of thumb. The classic study was carried out by two Oxford economists, Hall and Hitch, as long ago as 1939. They questioned the owners of thirty-eight firms and found that, rather than profit maximizing by producing where marginal cost is equal to marginal revenue, the majority in fact used cost-plus pricing. The entrepreneurs added up their costs of production and then added what they thought was a fair profit margin. A few took into account what the market price was, but none was able to calculate marginal costs and revenues.

Numerous studies since then have confirmed the validity of this pre-Second World War view of the world. For example, yet another such study has just been published in the *IBM Systems Journal*, which argues not only that 'Historical evidence suggests that cost-plus pricing has been in use at least since the end of the 18th century,' but that the principle is widely used in the pricing of services as well as products. Of course, there are many complicating factors, such as the price charged by competitors or the price which might be charged by competitors if your price is set at a particular level, but the basic finding remains valid.

The use of fairly simple rules to guide behaviour is in fact a rather sensible way to behave when confronted with a situation which is both enormously complicated and massively uncertain. The process of operating a business, far from being easy, is a very demanding activity.

The capacity of firms to deal with market situations in a cognitive sense, their capacity to process information and turn it into knowledge, is small compared to the sheer scale of the problems which confront them. Companies can never deal completely with the complexity of the real world. The uncertainty that shrouds the future is not so much a veil as an iron curtain. In the current state of scientific knowledge, it cannot be penetrated. There is ample opportunity at any point in time for any firm, no matter how large, to fail.

So far, we have looked at firms operating in the uncertainties of the market; at how firms have fared in practice, at the failure rates of the corporate giants of capitalism and at what, if anything, economic theory has to say about such failures. Governments are different. In the western world, the state has much more guaranteed continuity. Nevertheless, failure extends from the realm of economics into the world of public policy. It is to examples of failure of government policy that we now turn.

3 Up a Bit, Then Down a Bit

During the course of the twentieth century, western governments have become far more involved in the day-to-day lives of their citizens. Particularly during the final quarter or so of the century, the scope of the state increased markedly. In every western country, the size of state activity in the economy was considerably greater in 2000 than it was in 1950. In Britain, for example, in the middle of the twentieth century the most openly socialist administration in the history of the country was in power, but government spending as a percentage of the economy was lower then than it was under Mrs Thatcher in the 1980s.

By the turn of the century, almost 20 per cent of total national income in many countries in the European Union was diverted to subsidies for the poor, tax-break payments to households and other social benefits. Even in the United States, the comparable figure was over 10 per cent. In total, government spending accounted for as much as 48 per cent of national income in the EU and 36 per cent in the US. A great deal of this activity is designed to secure a greater degree of equality within the confines of the nation. Equality as an abstract concept is fraught by disputes and disagreement at the philosophical level. In the best tradition of political responses to difficult questions, I do not intend to deal directly with the 'true' meaning of equality. Instead, I adopt a much more pragmatic stance.

In the days of the universal franchise, the sense of governments that they need to do something about the status of the less advantaged parts of society is not merely driven by a sense of *noblesse oblige*, since everyone has a vote, and he or she can use it to turf out the government of the day at the next election. In the old Soviet Union, the party which purported to rule in the name of the industrial workers, and indeed whose leaders often mimicked their style of dress, was able to treat them with almost unparalleled contempt and brutality. Such an option is not available in the west. Governments must be seen to care. Yet despite the efforts of western governments for at least half a century, not only do social problems persist, but from time to time they appear to intensify. Society becomes less and not more equal. Social policies aimed at reducing inequality appear to fail.

One example of failure is provided by the experience of unemployment in Britain during the course of the twentieth century. Being out of work is the single most important determinant of poverty. This may seem so obviously true as to be hardly worth stating, but the popular press sometimes works itself into a frenzy about people living a life of unemployed luxury at the expense of the taxpayer. There are occasional examples of this, but in general people are much worse off when unemployed rather than in work, even when being paid just the minimum wage. This is not to say that some people might not prefer to be unemployed, but that is a different debate altogether. Unemployment remains a primary cause of poverty.

We might expect to find a relationship over the longer term between the size of the public sector in the economy and the rate of unemployment. A larger public sector might cushion the economy more effectively from the various shocks to which it is subjected from time to time and help preserve the level of employment in difficult times. This would not merely be due to the jobs themselves, but the spending of those with secure public sector jobs would help sustain jobs elsewhere in the economy. We do not mean in this context the short-term variations in public spending that might take place over the course of the booms and recessions of a single economic cycle. These rarely last for more than ten years, and most are considerably shorter. Rather, we are looking for evidence of the experience of an economy over the course of several decades, when the peculiarities of any individual cycle can be expected to be averaged out by their inclusion with others.

Britain, the first country to industrialize, has data describing the state of its economy which go well back into the nineteenth century. On average, over the years 1870 to 1938, just before the Second World War, the unemployment rate was 4.9 per cent.* This average conceals sharp fluctuations. At the peak of economic boom, for example, it fell to under 1 per cent, whilst in the depths of the Great Depression of the 1930s it rose as high as 14 per cent.

If we compare the period from 1946 to the present day with the period 1870–1938, we see that, on average, as a proportion of the economy as a whole, the public sector was well over twice as large. Yet the average unemployment rate from 1946 has been no less than 4.5 per cent. In other words, only very marginally lower than in the period 1870–1938, despite the massive rise in the importance of the public sector in the economy. And although the highest rate in any single year, at 11 per cent, was less than the 14 per cent of the 1930s, unemployment never fell below 1 per cent in the entire period since the Second World War.

The experience of unemployment in Britain during the sixty-odd years before and after the Second World War is very similar, both in its average rate and the range within which the rate fluctuated from year to year. Yet in the latter period the state was well over twice as large, relative to the economy as a whole, than in the previous period. Whatever benefits may have arisen from this massive increase in the role of the state, reducing unemployment, the primary cause of poverty, has not been one of them.

A more subtle but nevertheless very important dimension to equality is the concept of social mobility. It tells us about the opportunities available to individuals to better themselves over the course of their lives. A straightforward and widely used measure of social mobility is the proportion of children who remain in the same social group or class as their parents once they become adults, and the proportion who move out of it, either up or down the social scale. The higher the level of social mobility, the more likely it is that we will see children moving up and down between social groups.

* Excluding the First World War years, when unemployment was close to zero because millions were conscripted into the armed forces.

A corollary of high social mobility, when many children end up in a higher social category than their parents, is, of course, that there is an equal amount of downward mobility. Children of the well off and successful become more likely to end up in less desirable social classes. But a high level of social mobility is generally felt to be a good thing, since it implies that individuals are succeeding on their own merits rather than relying on the wealth and contacts of their families.

The *New York Times* in January 2003 reported on a study of social mobility in America. Sociologists Robert Perrucci and Earl Wysong analyzed the incomes and occupations of 2,749 fathers and sons between 1970 and the late 1990s. They concluded that social mobility had fallen sharply. As they became adults, children became more likely over this period to retain the class privileges or disadvantages of their parents. Studies in Britain also show a marked fall in social mobility over a similar timescale, from around the late 1960s to the present day. These outcomes cannot be ascribed in a simplistic way to 'Reaganism' or 'Thatcherism', because the trends were already established well before these often demonized figures ever came to power and have continued long after they lost it.

One factor widely seen as an important instrument to promote social mobility is education. Higher levels of spending on education, for example, are believed to promote social mobility, helping to unlock the potential of talented individuals from humble backgrounds. A particular illustration of this is reflected in the fierce debate that has been taking place in the last few years in Britain about access to the élite universities of Oxford and Cambridge. The fee-paying independent schools, confusingly known as 'public' schools in British English, educate a mere 7 per cent of the total number of eighteen-year-old school leavers in the UK, but almost one half of the Oxbridge intake each year is from these schools. Yet is this good or bad? Unlike schools in the state sector, the independent schools use academic selection rigorously to decide who can get into them in the first place. And what is the trend over time? Is the state sector share rising or falling?

Let's go back in time to the late 1960s, when I went for interview at what was shortly to become my college at Cambridge. As a keen sportsman, it was felt this was the college for me. Allegedly, as one entered the rooms of the Senior Tutor for interview, a rugby ball would be thrown. Candidates who caught it would be admitted to the college. Those who passed it back immediately in a flowing movement would be awarded a minor scholarship. And a major scholarship awaited those able to drop-kick the ball cleanly into the wastepaper basket. I have to confess that at that time it was 'as if' – using the economist's favourite phrase – the story were partly true. Many were admitted purely on academic merit, but sporting prowess still played a role.*

Degrees at Oxford and Cambridge are awarded in one of three categories depending on the level of intellectual prowess shown. For the better part of the twentieth century, however, Oxford retained what was known as the 'Gentleman's Fourth', an even lower category designed specifically for members of the landed gentry who could barely read or write. Yet admission in the 1960s was much more socially open than it was in the inter-war years. Both Oxford and Cambridge were very readily accessible to students from all backgrounds. My own was modest, my father servicing industrial heating systems for a living.† And there were many more just like me. Up until the late 1970s, pupils from the state system made up two thirds of all admissions to Oxbridge. An expensive independent-school education was becoming less and less of a way to secure access to the élite academic institutions.

In the past twenty-five years, the trend has moved sharply into reverse and it has become distinctly harder for a child of working-class parents to gain entry to Oxbridge. This is not in any way due to the admissions policies of the universities themselves. On the

---

* In fairness, the college now concentrates almost exclusively on intellectual ability, and over the past twenty years has been one of the best Cambridge colleges in terms of results in university exams. *Mais ou sont les neiges d'antan?*

† Though he served with great distinction in the Royal Air Force in the Second World War, winning the Distinguished Flying Medal for conspicuous bravery during night bombing operations over Germany.

contrary, they make far more effort than ever before to recruit people from as wide a social background as possible. But in spite of a massive increase in spending on state-school education since the 1970s, fewer and fewer of its pupils appear to be able to compete at the highest level and, by so doing, advance themselves dramatically both economically and socially. In this important respect, social mobility has fallen and policy has failed.

Social mobility tells us something about the dynamic nature of equality in a society, how it evolves over time. In addition, we can measure the degree of inequality directly at a point in time, taking a snapshot, as it were, of the outcome of the various trends in equality at a particular instance.

Rising inequality measured in this way has become a central concern of social scientists and politicians in the past twenty years, especially in the English-speaking world. It is not difficult to see why. From the end of the Second World War until the early 1980s, Britain and America enjoyed a long period of rapidly rising prosperity. This prosperity had been shared by all. The distribution of income was stable, with no marked trend either up or down. Then, in the early 1980s, there was a sharp change. Unemployment soared, reducing the incomes of the poor. As financial markets opened up and Wall Street and the City boomed, earnings differentials widened.

An influential cottage industry has developed, arguing that these trends are permanent features of western society. The articles almost write themselves, selecting from a list of phrases: 'globalization', 'social exclusion', 'the 40–30–30 society', 'life-long learning', 'the knowledge economy', and so on. The latest addition appears to be the 'diamond back', a configuration under which an élite 20 per cent are secure within the confines of a diamond-shaped fence, passing on their privileges to their children. The unfortunate 80 per cent live within their own diamond, and the two groups are connected by a very narrow isthmus, which only a very few ever manage to cross.

The most widely used measure of income inequality at a point in time is the so-called Gini coefficient. It is a method of summarizing how far the income distribution in a society differs from a completely equal one. By the way it is constructed the value of the Gini can only lie between zero and one hundred.* In a completely equal society, the Gini coefficient is zero – no inequality – and in a hypothetical society in which one person has all the income it is 100. So the higher the value, the more unequal the society. The exact details of its calculation need not concern us here, but essentially what it does is divide a society up into groups according to their share of total income and measure how much the result deviates from a completely equal society. The poorest 10 per cent of the population might, for example, have 3 per cent of its total income instead of the 10 per cent they would have in an equal society, the next poorest 10 per cent have 5 per cent of total income, and so on.

As with many concepts in the social sciences, translating it into practice is not straightforward. The economy and society are not physical systems which we can put on a pair of scales and measure exactly. We have to rely on estimates, with the result that different researchers can and do obtain different values of the Gini coefficient for the same country, even for the recent past. But, however it is measured, its value in Britain and America is now higher than it was twenty years ago. Deininger and Squire reported for the World Bank in 1996 some 693 calculations of Gini coefficients since the war in different countries around the world. For most of the post-war period, the US Gini coefficient was in the low to mid-30s; it now seems to be in the low to mid-40s. Britain is more egalitarian, with a Gini in the mid-20s for many years and now in the low to mid-30s.

The degree of inequality in the Anglo-Saxon societies, as measured at the close of the twentieth century, must be placed in a global context. Taking the world as a whole, the 1996 World Bank study found a minimum value of 18 and a maximum of 63, and offered the following generalization: 'Income inequality is much

---

* In some studies, the range is between 0 and 1 instead, but this merely requires multiplying every calculation by 100 to put it onto the more usual scale. The Gini measures the degree of inequality within a country, so, in terms of absolute levels of income, the poor in a country which is rich but unequal can be better off than most people in a poor but more equal society.

greater in Latin America and sub-Saharan Africa, which have Gini coefficients in the upper 40s, than in east and south Asia, which have Gini coefficients in the middle-to-upper 30s. The OECD countries, in general, have relatively egalitarian distributions of income, with Gini coefficients around 30.' In other words, most western countries are around 30, and values of 50 or more reveal wide disparities on the South American model. So, even after the impact of Mrs Thatcher, demonized by many on the left for her so-called 'divisive policies', Britain today does not stand out as a particularly unequal society by world standards. Indeed, the UK is now at the level which has existed in France, for example, for at least the last thirty years. UN estimates for France in the mid-1970s put the Gini in the mid-40s – similar to the US now. Despite French disdain for Anglo-Saxon capitalism, the signs of inequality in France itself are transparent, from the east European-style municipal blocks of apartments which surround the main cities, to high unemployment, to big discrepancies in regional income per head. Per capita income in the Île-de-France region, for example, is almost 90 per cent more than in nearby Nord-Pas-de-Calais.

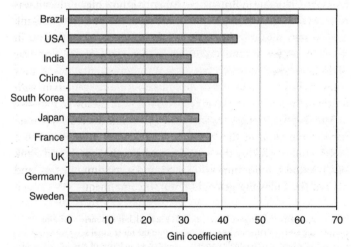

FIGURE 3.1 Estimates of inequality within various countries, mid-1990s. (*Source*: Deininger and Squire, World Bank)

From the 1996 World Bank study quoted above, we can extract the estimates made for some twenty western countries. The range of the Gini in developed countries is much narrower than that of the global data set. This suggests that there are stronger constraints on the degree of inequality in a developed democracy than there are in other societies. The single most frequently observed range of the Gini coefficient in the developed world is around 35 (which is exactly where Britain is). Values above 40 are unusual, suggesting that high degrees of inequality do not persist for long periods. There is, indeed, a powerful force containing inequality in the US at present: unemployment fell sharply during the 1990s and, despite the recession in 2001, is still low in many areas. So most people who want a job can get one; and most people are much better off with a job, even at the minimum wage, than they are on welfare benefits. The degree of inequality in the US now is no greater than it was in the early 1990s. It is still high by the standards of western democracies, but it is not getting worse, and by some accounts it is actually falling slightly.

Collecting data to estimate the Gini over long periods of time is extremely difficult. Estimates in recent years, for example, depend a great deal on the information available in the records of the income-tax authorities. Most adults in the west now either pay tax on their income or receive income in the form of benefits from the state. Indeed, given the massive complications of most tax and benefit systems, many people both pay tax and receive benefits in one form or another. The point here is that reasonably reliable information now exists on the individual incomes of almost all adults. For most of history, and even for the history of the last two hundred years of industrial capitalism, this was not the case. State spending as a proportion of the overall economy was much lower, and so most people were not required to pay income tax.

More reliable estimates can be made of the share of national income going to the very wealthy, for whom tax information does exist. Entirely separate procedures have been developed for estimating the total size of the economy, without relying on data on individual incomes. Simon Kuznets, one of the first Nobel prize-winners in economics, pioneered many important methods of collecting and estimating data in economics, including ways of thinking about the total income of an economy. In 1953, he published a study of the share of income of the very wealthy in the US in the first half of the twentieth century. Up until the mid-1930s, the top 1 per cent of all earners took between 12 and 16 per cent of all income. The precise figure fluctuated from year to year, sometimes sharply, but it stayed within this range. Over the next fifteen years, to the middle of the century, there was a dramatic fall to around 8 per cent. But the fall itself was not smooth, with year on year movements differing substantially.

So, over the first half of the previous century, income distribution within the US appears to have become much more equal. But the movement towards equality was erratic, with some short periods showing large changes, and others showing hardly any at all and indeed sometimes showing an adverse movement against the longer term trend. In the second half of the century, we seem to have had a long period when the income distribution was both rather egalitarian and stable, followed by a fairly short period when the Gini coefficient rose and inequality increased, and then a decade or so of relative stability once again.

In other words, the distribution of income within countries is fluid and dynamic, sometimes appearing to be stable, then moving suddenly in one direction or the other. There is a tendency over a long period of time for income to be more equitably distributed, but around this trend there are irregular fluctuations of various sizes and lengths of time.

What sort of patterns, if any, emerge when we look at a broader picture of inequality? What do the data tell us when we shift our focus from movements in income inequality within developed economies to one of average incomes in different countries around the world?

Again, if the Jeremiahs are to be believed, the picture here is one of unrelieved gloom and despondency. Giant multinational corporations roam the globe, every bit as ferocious a set of predators as the dinosaurs of old. Poor countries are stripped of their resources, and the gap between rich and poor grows wider and wider. The policies which encourage this sort of behaviour are alleged to be creating wider inequalities on a world scale.

We might reflect back on the previous chapter, and the formation of companies on a global scale around a hundred years ago. It was the early twentieth, not the early twenty-first, century when globalization first became a reality, and the ability of the developed world to exploit the rest, if indeed this is what happens, was even greater. The corporate giants of a century ago could rely not just on contracts and trade agreements, but often on direct political control of many areas of the world, for colonization was at its height.

The current-day pessimists are completely wrong when they repeatedly assert that the world as a whole is becoming more unequal. For the first time in the history of capitalism, the distribution of world income between countries became more equal in the second half of the twentieth century, and in particular during its final quarter. The economic successes in Asia have liberated many, many millions of people from lives of unremitting drudgery and toil, and have reduced world income inequality. But, just as with the movements over time in inequality within nations, changes in global inequality have not happened smoothly.

Before looking in more detail at the current distribution of world income we need to go back in time. Estimating Gini coefficients for the early nineteenth century is not easy, but the most thorough research on national incomes around the world over long periods of time has been carried out by Angus Maddison. Using the data from his monumental work *Monitoring the World Economy 1820–1992*, we can observe very clear trends in the degree of equality of world income.

In the early nineteenth century, average income per head across countries was compressed in a very narrow band: there was a kind of equality of misery. Only a few countries in northern Europe and America had begun to industrialize; the world Gini coefficient appears to have been in the low teens. So a small number of countries were beginning to escape the poverty and hunger which had

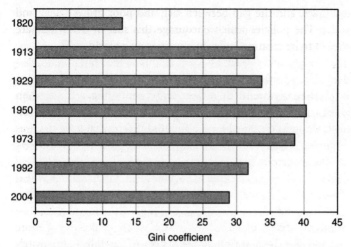

FIGURE 3.2 World income inequality between countries 1820–2004.
(*Source*: Maddison (1995) and estimates from official data for 2004)

been the lot of most of humanity since time immemorial. By present-day standards, even these countries had remarkably low standards of living, but the vast bulk of humanity remained even poorer.

The subsequent dramatic success of capitalism in Europe and its offshoots in North America and Australasia led to a marked widening of the degree of world inequality. The rest of the world appears to have benefited from the stupendous rise in prosperity in the west, but by comparison progress elsewhere was slow. So, by the middle of the twentieth century, the world Gini coefficient stood at 40.

The club of prosperous nations that had formed by about 1870 had hardly changed by 1950. Scandinavia was an exception as a late entrant, but membership appeared to be fixed. Following its humiliation by the west in the second half of the nineteenth century, Japan had embarked on a programme of modernization, but at the outbreak of the Second World War it had by no means caught up. Japanese growth in the 1950s and 1960s was phenomenal, and other, much poorer countries in east Asia gradually began

47

to experience rapid growth in living standards. The net result was a sharp fall in global inequality by the early 1990s. Since then Japan has faltered, but the rest of east Asia, despite the crisis of 1998, has continued to grow. In particular, now that India and China have moved away from the disastrous institutional structure of a planned economy, their prosperity has improved noticeably. Africa, of course, remains in very serious difficulties. In terms of world inequality, however, events there are dwarfed by developments in Asia. The total population of Africa is well under 1 billion, whereas Asia has a total approaching 4 billion. GDP per head in real terms in Africa as a whole is probably lower now than it was in the late 1960s; over the same period in Asia as a whole – including the two giants, China and India – the figure has almost tripled.

The recent changes in inequality within countries such as South Korea and China mirror the evolution of global inequality over the past two centuries. In the 1950s, South Korea was a poor but very egalitarian society. During its phase of phenomenal growth, inequalities opened up and its Gini coefficient approached 40. Now that South Korean income is close to average income levels in the west, prosperity is being more widely shared, and the Gini has dropped back to the low 30s. China is at a much earlier phase of development, and estimates for this country are more than usually unreliable, but it does seem as though the boom in parts of China in recent years is leading to marked increases in inequality.

By far the most important characteristic of capitalist economies, which distinguishes them from all other previously and currently existing societies, is their slow but steady underlying rate of real economic growth. Before the nineteenth century, increases in living standards of just one percentage point took decades or even centuries to achieve. Since then, per capita growth of around 2 per cent per year has become the norm in the successful market economies.

The impact of moving to the path of sustained growth can be clearly illustrated by a comparison of per capita incomes in South Korea and an African economy. The Ivory Coast is by no means unsuccessful in African terms, but the transformation of South

Korea has been stupendous. Even as late as 1970, the two countries had very similar levels of output per head, with Korea being about one third higher. But it is now nearly nine times higher.

Predicting future trends in global inequality is very difficult. To some extent, this depends on the outcome of events in eastern Europe. If the former Soviet bloc begins to succeed economically, further downward pressure would be exerted on world inequality. Events in Latin America, too, send mixed signals about future prospects.

Yet this is intellectual speculation, mere 'ifs' and 'buts' compared to the certainty of what has actually happened. A group of Asian economies has grown rapidly in the closing decades of the twentieth century, and world inequality has been reduced. In China and India, average real incomes have approximately doubled in the past thirty years – considerably more in China, according to some estimates. Indonesia, with over 200 million people, has an average income similar to that of France and Germany in the early years of the twentieth century – not rich by contemporary standards, but much, much higher than in most African countries – while South Korea and Taiwan are close to current European levels. Further, once a country has managed to create the elusive mix of institutions and motivations which deliver a successful economy, it does not tend to fall back into general poverty.*

Movements in world inequality mirror developments within individual countries. This is not to say that their precise paths over time are the same – far from it – but the underlying characteristics are very similar. We see fluctuations of different lengths and sizes around apparently well-established trends, and in the case of world inequality, the reversal of a seemingly inexorable trend towards wider inequality which had persisted for over 150 years.

We might reasonably ask to what extent it is possible to control a concept that moves so erratically. In fact, attempts to anticipate its changes and hence plan to control them are inherently flawed.

---

* Argentina is a possible exception to this rule. Before the Second World War, the country had European levels of prosperity, but its own disastrous subsequent policies held it back.

Failure is endemic to any such policies. The difficulties of assessing future movements in the degree of inequality are intensified by the lack of any theoretical framework in which to work. A general theory which enables systematically accurate predictions to be made of the movements in inequality is lacking in the standard approaches in the social sciences.

In economics, the central intellectual construct of conventional theory is the so-called general-equilibrium theory. This describes a static world in which the amount of resources available is fixed and in which everyone's tastes and preferences are fixed. The task is to allocate these in an efficient way. The work of allocation is done entirely by the price mechanism. In a single, simple market, such as that for wheat or pigs, say, we can readily imagine the price adjusting so as to bring supply and demand into balance. General-equilibrium theory describes a competitive economy in which a set of prices exists at which supply and demand balance out in every single market. In this Platonic idea of a market, unemployment is zero and all is well with the world.

There are many criticisms of the concept of general equilibrium. Some are empirical, noting the huge gulf between the world described in the theory and the world which appears to actually exist. The most powerful are themselves theoretical. The findings of orthodox mathematical economists themselves have done more than anything to show the serious limitations on the usefulness of general-equilibrium theory.* Just one example will serve as an illustration. The Stanford economist and Nobel prize-winner Kenneth Arrow's contribution to modern economics has been enormous, even though his name is little known outside the world of high theory. Despite having done more than anyone to place general-equilibrium theory on a modern mathematical basis, he is on record as saying that he regards the theory as being empirically refuted – a very strong statement from a scientist. If a theory is refuted, it is wrong.

---

* An extended discussion of many of these is provided in English in one of my previous books, *The Death of Economics*, Faber and Faber, 1994.

For our present purposes, however, the most relevant feature of general-equilibrium theory is that it tells us nothing at all about the degree of inequality in a society. The theory allows *any* degree of inequality to emerge as a solution to its equations. It is only concerned to find a set of prices which will in principle allow supply and demand to balance in all markets. It is not concerned about whether those prices lead to a fair or to a dramatically unequal society. Some of the prices required to be established are, for example, the prices for various kinds of labour, which is, of course, purchased by firms. The price of labour is more usually known as a wage or salary. General-equilibrium theory, the core achievement of conventional economics, is simply not concerned about whether wages are high or whether they are at near-starvation levels. It is simply concerned to find a set of prices, including that of labour, at which supply and demand are in balance everywhere.

In so far as free-market theory considers the concept of equality, it focuses on the so-called Pareto concept. This refers to a state of affairs in which no person can be made better off without someone else being worse off. Enormous intellectual effort has been spent in establishing the theoretical conditions under which this so-called Pareto optimum might obtain.* But in the context of most discussions of inequality, the concept itself seems rather bizarre. For example, almost everyone would agree that if Adolf Hitler had been assassinated in 1933 just before he came to power in Germany, the world would have been a much better place. We might well have been spared both the Second World War and the mass exterminations under the Nazis. Yet, according to the Pareto concept of equality, this would not have been a good thing at all because at least one person, in this case the murdered Hitler, would have been worse off. The fact that many millions would have been better off counts for nothing according to the Pareto concept. Indeed, the very fact that one person would have been worse off

* Purely out of interest, for any sad people such as the present author who are fascinated by these things, a general equilibrium in a timeless single period world *is* a Pareto optimum but, once the future is allowed to exist (!), in general it is not.

means that the outcome would, according to the Pareto concept, be *less* desirable, no matter who that person might be.

From a completely different perspective to the general-equilibrium theory of free-market economics, we have Karl Marx. The concept of the immiseration in absolute terms of the working class was put forward in *The Communist Manifesto* in 1848. Marx believed that his iron law of wages would ensure that the workers would be kept permanently at a standard of living just about sufficient to remain alive. This has been thoroughly discredited by events, as we have already seen. By the standards of today, even many of the mid-nineteenth-century bourgeoisie were paupers. During the course of his life Marx himself noted difficulties with this hypothesis and postulated instead that capitalism would result in workers becoming poorer only relative to capitalists. Again, this has not happened. The theory has been refuted empirically. The share of wages in national income is considerably higher than it was a century ago, and the share of profits smaller. Workers have become better off relative to capitalists. Indeed, we can go even further than that, for most workers have become in part capitalists themselves. Pension schemes, for example, receive much of their income from dividends on equities, which are, of course, paid out of the profits generated by companies.

Yet another thesis on equality is provided by the theory of economic growth. The focus of this is on the distribution of income between countries over time. Nobel prize-winner Robert Solow formulated this in a seminal article written almost fifty years ago. A central prediction of this theory is that, in the long run, average per capita incomes between all countries will converge. An important recent refinement of this model, based on the superbly named 'post-neo-classical endogenous growth theory', implies that this will not necessarily be the case. Instead, countries will be divided into groups, and convergence to the same level of income will take place within, but not necessarily between, individual groups.

Using post-war data from across the world, economists are unable to decide which of the two versions of growth theory describes the world better – and a dispassionate observer might conclude that neither is very successful by scientific standards. But the problems for both these theories are illustrated nicely by the convergence, or rather by the lack of it, in the average income levels in the individual states of America. This evidence is documented in detail in an article written in June 2001 by G. Andrew Bernat in the *Survey of Current Business*. Many of the practical factors that might prevent convergence between different countries – such as tariff barriers which deter access to wealthy markets, the lack of a settled and enforceable system of law which makes it hard for business to develop – are essentially absent when we compare average living standards across the US. Yet in 2000, for example, income per head in the richest US state, Connecticut, was almost twice as high as in the poorest, Mississippi.

A theory which predicts that in the long run all areas will converge on the same level of income per head is clearly in serious difficulty if this is not actually observed even for areas within the same country. The same sort of gap that we see between states in America exists between the individual countries of western Europe and, for the most part, between the regions within each of these countries.

In short, the mainstream social sciences* have not succeeded in establishing any firm theoretical guidelines on the evolution of inequality over time, whether between the individuals within a country or between countries themselves. Further, the empirical evidence seems to indicate that inequality in both these concepts moves in irregular ways which are difficult to predict.

There is an important reason that explains both why it is hard to construct a successful theory from conventional mindsets in the social sciences and why we see the kinds of patterns in inequality that we do: the degree of inequality in any given society is the outcome of the interactions of millions of individual agents, both people and companies. Their tastes and preferences are not fixed

---

* Alas, one must include Marxism, for this empirically falsified doctrine nevertheless retains many devotees even today, particularly within the social science and arts faculties of western universities.

but depend to varying degrees on the behaviour of others. Individuals and firms copy and adapt their behaviour over time. Institutions alter and change. Grand themes may appear to determine the outcome for many years into the future, but the reality is far more complex. Too many factors can influence the degree of equality for it to follow a well-understood, deterministic path.

The dimensions of the problem that confronts policy-makers in trying to decide strategy, with many different factors affecting the outcome of any decision, are too large to be grasped in their entirety. In these circumstances, there is ample scope for decisions to go wrong. In short, there is ample scope for strategy to fail at any point in time.

The curse of dimensionality, as we might call it, affects attempts to control the movements in inequality. The same concept helps to explain why so many of the big developments in the past fifty years were not anticipated. The 1930s saw millions unemployed, but the post-war decades experienced rates of unemployment lower than the wildest dreams of the Keynesian enthusiasts of the 1940s. As late as 1979, conventional wisdom would have dismissed the idea that unemployment might rise in the UK to over three million and would have been even more sceptical about the government under which this happened getting re-elected. In the early 1990s, very few of us imagined that full employment would return to large parts of Britain by the end of the decade. On a more grandiose scale, the sudden collapse of the Soviet bloc came as a surprise even to free-market ideologues. The rise of the east Asian economies was equally sudden and dramatic.

In short, we cannot anticipate how the degree of inequality will move over the next, say, ten years, either within a single country or across the world. Like many phenomena in the social sciences, inequality appears to have the characteristics of what is known as a complex system. There is an inherent lack of predictability. Whatever the method used, it is not possible to make consistently accurate forecasts. The distribution of outcomes over time may be reasonably stable, so that we can make meaningful statements about the range which the system will explore, but we cannot say with confidence where it will be at any given moment.

An analogy might help to illuminate the issue. We might think of the problem as, say, attempting to decide whether a particular radio signal received contains genuine information (for example, a piece of music) or is simply a combination of random squeals and hisses. Modern mathematical techniques can identify the proportion of the sequence that contained recognizable patterns and which could therefore be presumed to be genuine music (the 'signal'), and the proportion which was simply interference (the 'noise'). The existence of a relatively high ratio of 'signal' to 'noise' is a necessary condition for reasonable forecasts of the relevant data series to be made. A series dominated by noise is very similar to a purely random series which, by definition, cannot be consistently forecast with any degree of accuracy.

Analysis of, say, annual post-war data on inequality in the US or year-by-year movements in world inequality gives very clear results. The data appear to contain a small amount of information, but they are actually dominated by noise rather than signal. The implications for predictions of the degree of inequality are clear: no matter what *ex post* rationalization is given for the past and no matter how it is derived, it is simply not possible to make accurate forecasts about the future.

The one thing of which we can be certain in trying to predict and control the degree of inequality is that we cannot be certain of the outcome. In order to control a system – any system, whether an economy, a biological system or a machine – we need to be able to do two things: first, make forecasts which are reasonably accurate in a systematic way over time; and second, understand with reasonable accuracy the effect of changes in policy on the system one is trying to control.

Unless policy-makers know with reasonable confidence what state the system is likely to be in at some point in the future, it is not possible to say what action is required now in order to bring about a more desirable outcome. And unless the authorities understand the impact of their actions, it is not possible to know what should be done in order to bring about any desired outcome. The scope for failure abounds.

Nevertheless, many people continue to believe that, in order to design effective policies, all we need to do is collect more information and statistics, analyze the data and produce a plan which will solve whatever problem confronts us. But there are deep underlying reasons for the inability to plan and control outcomes successfully. We have discussed one of them in this chapter, using different aspects of inequality to illustrate the point. The historical data which we have is dominated by noise rather than by signal and contains very little true information. No matter how hard we try, no matter how many statistics we collect, there are strict limits to the value of the genuine information we can extract.

The persistent failure of economic forecasts is perhaps a more familiar example of this point. Regardless of the data which is gathered, the statistical technique which is applied or the particular economic theory which is used, economic forecasts over time are bound to contain substantial errors.* The economy at the aggregate level behaves much more like a purely random system than one which can be predicted and controlled.

It may seem implausible that economic systems behave as if they were almost random. However, this near-random quality does not mean in any way that the individual components of an economy – people, firms, governments – take decisions at random. On the contrary, they act with purpose and intent. But the consequences of these millions upon millions of individual decisions, interacting with each other all the time, lead to an overall outcome, for total output (GDP), say, that appears as if it were close to being random. The sheer dimensions of the problem are simply too great for the system to be understood properly. There are simply too many factors that determine the outcome, and whose relative importance alters over time, for the complete picture ever to be grasped.

* For a technical exposition of this, see, for example, my paper with Craig Mounfield, 'Random Matrix Theory and the Failure of Macro-economic Forecasting', *Physica A*, Vol. 280 (2000).

There are often even more fundamental limitations to the ability to predict, plan and control social and economic systems, a problem which can arise in situations which are not complicated, but simple. And it can arise when, far from the available information being in some way incomplete, a great deal is known about the problem in hand. Indeed, even when the rules of behaviour of the individual agents are transparent and known to all, and when any single move made by any single agent can be observed readily by all the participants, the precise outcome for the system as a whole can still be unpredictable. In other words, even when we have all the data and all the information that exists in a particular context, uncertainty can still prevail.

This is the theme introduced in the next chapter, which we illustrate with the topic of segregation in residential neighbourhoods. And it is a theme that will emerge time and again in the rest of the book, each time in the context of a specific practical problem relating either to business or economic and social issues.

4 Making Sense of Segregation

Friedrich Engels was indirectly responsible for some of the greatest crimes against humanity ever to have been committed. The son of a prosperous industrialist, he was born in Germany in 1820. When he was a young man, his father sent him to manage his cotton factory in England. The city where it was situated, Manchester, was the very centre of the Industrial Revolution, which had sprung upon the world some fifty or so years previously.

Engels was shocked by the poverty which he encountered in the city. Like many guilt-ridden members of the bourgeoisie, he was determined to do something about what he saw as an affront to civilization. He simply could not have imagined the consequences of one of his actions. He began communicating his distress in articles for a radical journal called *Franco-German Annals*, edited by one Karl Marx. The two men met in 1844 and soon became close friends. Engels continued to draw money from the family business and used it to support Marx throughout the rest of his life. Whenever the two men were separated, Engels used to send Marx two separate letters virtually every other day, each one of them containing a piece of a £5 note cut in half.

Marx was perfectly happy to live off the profits of capitalism, at the same time as he denounced it fervently in *The Communist Manifesto* (cowritten with Engels) and *Das Kapital*. The inspiration of the Soviet and Chinese communist parties, Marxism led to the enslavery and impoverishment of hundreds if not thousands of millions of people, and to the systematic murder of tens of millions of innocents.

Here indeed is a dramatic example of a policy which failed, whose outcomes were spectacularly different from those which the protagonists intended. Engels and Marx had a strategy which they thought would help to alleviate the sufferings of the working class in the early decades of industrial capitalism. Instead, they helped to create a monster.

As well as working in the family business, Engels was something of an entrepreneur. He was, for example, a successful author in his own right, and his initial gifts to Marx came from the profits he made on his first book. This, *Condition of the Working Classes in England*, was published in 1844 and has since become a classic. Although interspersed with tedious political diatribes, the book offers a careful and detailed description of the living conditions at the time of the workers in Manchester and its satellite industrial towns. One might imagine a modern-day version being written by, say, a reporter on the *New York Times* or the British *Guardian*.*

There is no doubt at all that life at the time was exceptionally hard. Engels' description of everyday life amongst the industrial workers is an accurate one. Even though Britain was then the richest country in the world in terms of average income per head, by present-day standards the country was very poor. Comparing living standards over such a long period of time is difficult, but in terms of present-day purchasing power, incomes in the UK in 1844 were perhaps one tenth of their current levels. We can think of England at the time as a modern-day Bolivia.

A striking feature of Engels' Manchester was the very marked geographical segregation that existed between rich and poor. The words of the author himself describe it best: 'A person may live in Manchester for years, and go in and out daily without coming into contact with a working people's quarter or even with workers, that is, so long as he confines himself to his business or pleasure walks. This arises chiefly from the fact, that by unconscious tacit agreement ... the working people's quarters are sharply separated from the sections of the city reserved for the middle class.'

Rather unnervingly, Engels goes on to offer a description of the very centre of the city which still rings true today: 'Manchester

---

* Indeed, after the first draft of this chapter was written, at least one such book *did* appear, but I make no claims to be able to predict the future in any more general sense!

contains, at its heart, a rather extended commercial district, perhaps half a mile long and about as broad, and consisting almost wholly of offices and warehouses. Nearly the whole district is abandoned by dwellers and is lonely and deserted at night … this district is cut through by main thoroughfares upon which the vast traffic concentrates, and in which the ground level is lined with brilliant shops. With the exception of this commercial district, all Manchester proper is unmixed working-people's quarters, stretching like a girdle averaging a mile and a half in breadth around the commercial district. Outside, beyond this girdle, lives the upper and middle class.'

What has changed over the past 150 and more years in England's second most important city? The commercial centre is still small, not much larger than in Engels' time, and does indeed contain offices, 'brilliant shops' and 'vast traffic'. Until very recently, when old warehouses began to be converted into young urban professional apartments, the centre of the city was still deserted during the night, once the pubs and restaurants had closed, for no one lived there. The thickness of the girdle of 'working-people's quarters' has widened, but it still surrounds the city. Within a fifteen-minute walk of the financial services district of Manchester, one can be in some of the poorest areas in western Europe. Instead of the incredibly cramped, dirty terraced houses described by Engels, we see enormous, decaying tower blocks, reminiscent of East Berlin, Moscow and Warsaw.

In short, the geographical segregation along class lines within the city of Manchester is very similar in 2004 to that described by Engels in 1844. There are marginal changes here and there, with some districts going up in the world relative to others and some going down. But his social map of the centre of Manchester and its immediate environs is at once recognizable today. A tourist armed with Engels' guide would succeed in avoiding many of the worst and potentially most dangerous areas.

But how can this be? Generations of social reformers, local authorities and governments have worked assiduously at promoting the well-being of citizens, at promoting a more integrated and cohesive society. And yet, in Manchester in the opening years of the twenty-first century, we see a pattern of social segregation which has existed since the early days of Queen Victoria. The 'unconscious tacit agreement', the process that separates social groups, persists. Almost all attempts to break down segregation appear to have failed.

The same phenomenon can be seen in various guises throughout the cities of western Europe, which have long been segregated along class and income lines. The major exception is inner London, where the rich and the poor have traditionally lived cheek-by-jowl. But here the rich spend a great deal of effort and money in insulating themselves from their immediate surroundings, relying on private education, private health care and private transport. In contrast, Paris is much more typical. The centre of the city itself is very affluent, with both the indigenous poor and immigrants being dumped outside the city's main ring road, separated not merely geographically but in political boundaries as well from the vibrancy and energy of the main city.

Many similar examples can be found. The housing division along religious lines, dividing Protestants and Catholics, in Ulster is unusual in the west nowadays. More common is the segmentation within the working class between the respectable and the remainder. The former have gradually embraced values more traditionally seen as being associated with the middle class. The areas in which they have chosen to live symbolizes their progress, having fled their industrial slums, first to rented accommodation in garden estates and then to private ownership. This process began in many European countries in the 1950s and continues to this day. Despite all attempts to promote social integration, segregation persists. Policies fail.

British society received a number of shocks in the summer of 2001, when extensive urban rioting took place in the north of England, involving white youths and those from the hitherto quiescent Asian community. One of the biggest shocks, particularly for metropolitan liberal thinkers, was the subsequent realization that many of England's poor, northern towns and cities are segmented along racial lines. Much in the same way as King Edward VIII organized an expedition to the poverty-stricken coal-fields of south Wales during the economic depression of the 1930s, distinguished London-based committees have visited, discovered and pronounced. It appears that, far from celebrating diversity, the masses have taken every opportunity to separate themselves along racial lines. Again, the 'unconscious tacit agreement' of Engels seems to be prevailing over well-intentioned policies to promote integration.

The geographical separation of races is, of course, a major feature of American life. A dramatic illustration is given in Tom Wolfe's novel *A Man in Full*, a complex and interwoven tale set in contemporary Atlanta, Georgia. The black mayor of the city has arranged to be driven around in order to illustrate exactly these divisions to his old fraternity classmate. 'The drive went under the Highway 75 overpass and then past the old Atlanta Convention Centre, which meant they were already on Piedmont. Then the Mayor said, "You know what street we just crossed? Ponce de Leon." This required no amplification, since practically everybody in Atlanta old enough to care about these things knew that Ponce de Leon was the avenue that divided black from white on the east side of the town ... They might as well have painted a double line in the middle of Ponce de Leon and made it official, a white line on the north side and a black line on the south.'

Immense efforts have been made in the US over many years to promote integration, yet the divisions along racial lines remain sharp. *The Economist* magazine reported in February 2003 that American schools were more segmented on racial lines than they were in the 1960s. And it was precisely in the 1960s that major initiatives began to be taken to improve integration and break down the old racial barriers. Participation in the peaceful demonstrations in the Old South against segregation remains a badge of honour amongst American liberals of a certain age, many of whom now occupy prominent positions in politics and the professions. The federal government sent the National Guard to ensure the integration of schools. Spending on health and education, much of it specifically targeted at minority communities, rose dramatically. Yet, despite all this, segregation remains. Policies have failed.

The separation extends beyond geographical and residential confines. To take just one example, there are dramatic differences in the distribution of income across groups. The poverty rate across America amongst non-Hispanic whites is some 8 per cent, but the comparable figure for blacks is 24 per cent. Non-Hispanic whites make up around 70 per cent of the total US population but considerably less than one half of the total number in poverty. The highly productive and innovative Harvard economist Ed Glaeser has documented a wide range of other areas that illustrate this divide. He finds, for example, that in America people of different races are more likely to cheat on each other and that the degree of racial heterogeneity of an area is linked to the probability that a riot will take place.

Economists have long been aware of the phenomenon of persistent segregation of American society along racial lines. Nobel prize-winner Gary Becker was the first to devote serious theoretical attention to the problem. His seminal work, *The Economics of Discrimination*, was written as long ago as 1957 and has generated a large literature within conventional economics.

Becker went on to bring other important social issues, such as marriage and crime, under the scrutiny of economic theory. Some may feel that this is yet another illustration of Stephen Potter's description of the ubiquitous nature of the applications of economics: 'It is generally agreed that almost any phrase from any chapter of this extraordinary subject will meet any emergency, if the sentences are spelt out sufficiently clearly and slowly.'

The work of Becker cannot, however, be ignored in any serious consideration of these social topics, for he used the central insight of economics, that agents respond to incentives, for the first time in these areas of social concern. This is as near to a universal law of behaviour as we have in the entire set of social sciences. For example, in the UK the authorities have installed large numbers of cameras to monitor the speeds of vehicles on various roads.

Drivers exceeding the speed limit attract an automatic fine. It is not certain that a speeding motorist will be punished in this way, because the film for the camera is costly and the cameras are often left empty. Nevertheless, motorists who are exceeding the speed limit almost invariably slow down whenever they realize they are approaching one of these cameras.

Many of the other disciplines in the social sciences tend to be dismissive of the insistence on the importance of incentives in economics. However, I have noticed that whenever I have been driven by, say, a sociologist or a lawyer, they too slow down when they encounter a speed camera. Incentives matter.

Economics does have something useful to say about social issues such as segregation, crime and the family. But economists go on to weaken, and sometimes even spoil, their own case by insisting that agents – people, firms, governments – always act according to a rational calculation of their own self-interest. In economic theory, individuals are assumed to be able to process vast amounts of information and to carry out complicated mathematical analysis in order to maximize their own 'utility', in the jargon of the discipline. In other words, individuals are presumed to be self-interested, anxious only to further their own well-being. Yet there does appear to be something inherently implausible, for example, about the idea that young working-class men assess all the available information and choose the 'optimal' decision when they are contemplating breaking into a car or thinking about punching someone in a dispute in a bar.

So, for conventional economists, even those as imaginative as Becker, segmentation along marked racial lines can only persist if people of one race actively dislike people of another. People have their own tastes and take decisions which are best for them, given their preferences.

On this view of the world, segmentation of neighbourhoods along racial lines can only persist if people have very definite preferences for this particular outcome. Their tastes are such that they gain substantial 'utility' from living in an area where most people are the same as them. Conventional economic theory does not attempt to explain why a person's tastes are what they are. It takes them as given and fixed. Equipped with these tastes and a certain level of income, the individual then chooses the products and services to consume which will bring him or her the most satisfaction – or, again, 'utility' in economic terminology.

It is self-evident that an individual, with the possible exception of a tiny number of stupendously wealthy ones like Bill Gates, cannot obtain absolutely everything he or she wants. As the Rolling Stones once sang, 'You can't always get what you want,' or, as an economist would put it, individuals maximize utility subject to constraints. In particular, people have limited incomes, so they are obliged to choose between alternatives. Further, different individuals derive different amounts of 'utility' from the same set of available products and will in general make different choices. Most people who prefer Pepsi to Coke, or vice versa, rarely switch between the two.*

One 'product' that is available is the amount of satisfaction, or utility, which derives from the racial composition of one's neighbourhood. It is only one of many different types of utility which flow from the choice of dwelling. Some people, for example, prefer apartments to houses, others like the bustle of the crowded city, others prefer the suburbs. All these factors provide their own separate amount of utility to the individual exercising a particular choice. The utility which comes from the racial composition of the area is just one amongst many.

So, when we observe very marked segregation in residential areas along racial lines, the implication of economic theory is that individuals must have a very strong preference for racially homogenous neighbourhoods. A large proportion of the population must each get a large amount of utility from living in a segregated neighbourhood, on a scale sufficient to outweigh the utility derived from the other aspects of the nature and location of their home.

The theory seems very plausible at first sight. We observe

* Except, of course, during the great debacle of the 1980s when Coca Cola altered its formula. *There* was a policy that failed spectacularly! But Coke recovered in time to prevent complete disaster overtaking them.

marked segregation across cities and states, and so it may seem reasonable to presume that individuals have marked preferences for segregation. The system at the overall level exhibits segregation, so surely this must arise from the preferences of individuals. Unfortunately for the theory, the empirical evidence does not give much support to the view that individuals have widespread, strong racial preferences. Indeed, although a very small minority hold these opinions, across the population as a whole there is no evidence that this is the case. There is evidence in the US of a certain amount of preference in favour of people in the same racial group, but it is not at all overwhelming. Indeed, two intriguing academic analyses of the popular television game *The Weakest Link* found discrimination very hard to detect.

Contestants are required to answer general-knowledge questions, building up a shared pot of money for correct answers. At the end of each round, one contestant – deemed to be the 'weakest link' – is voted off by the others. In the final round, the two remaining contestants compete for the whole of the prize money. The person who is voted off does not by any means have to be the weakest competitor in any objective sense. Indeed, it is often tempting to eliminate a rival who appears to be good at answering the questions rather than the one who has answered the least questions or who has given the highest number of incorrect answers.

By the end of 2002, more than 1,000 contestants had taken part in the show in the US. Kate Antonovics, Peter Arcidiacono and Randall Walsh examined who people voted to expel, taking precise account of factors such as how many questions a contestant had answered correctly and the votes cast by other players. They could find no evidence at all of discrimination by whites against African Americans. Steve Levitt looked at the kind of voting strategy that people followed. In the early rounds, it makes sense to vote off players who really are weak, so that the total value of the pot can increase faster. But soon it is the strongest player who becomes most vulnerable, since everyone else wants him or her out of the game. Again, he found no evidence of racial discrimination.

In a broader context, the General Social Survey covers between

1,200 and 2,400 individuals every year since 1972 in the US and provides valuable information on attitudes to social issues. Ed Glaeser and his colleagues Alberto Alesini and Bruce Sacerdote have looked at this in some detail. The Harvard researchers examined the question 'Do you think the state should spend more on welfare?', and used standard statistical techniques* to analyze the data. These techniques, it should be said, are able to distinguish the impact of individual factors, even when these factors themselves tend to be related to each other.

The scholars found a number of interesting results. The main point in this particular context, however, is that there is some evidence of racial preference, but it is not strong. According to the analysis of the survey, blacks are 23 per cent more likely to say that welfare should be increased compared to non-blacks. A similar effect is discovered when the General Social Survey asks whether there is support for government redistribution of income. However, analysis of the responses given by whites only suggest that, overall, differences are modest. For example, if distinct prejudice exists, whites in states with a higher percentage of blacks should be less likely to support welfare spending than states with a lower percentage, yet there is no difference in the attitude of whites along these lines at all.

A certain amount of racial preference is found in a different piece of analysis carried out by the same researchers. Individual states in America have considerable discretion in the generosity of payments in the Aid to Families with Dependent Children (AFDC) scheme. In every single state, blacks are both in a minority of the population as a whole and at the same time are disproportionately represented amongst the poor. The maximum AFDC payment to a family of three tends to be lower the higher the per-

---

* The basic principle underlying these techniques was discovered more or less in an afternoon some 200 years ago by the mathematical genius Friedrich Gauss, one of the three greatest mathematicians in history. Since then, numerous layers of sophistication have been built on the original discovery, to the extent that a (well-deserved) Nobel prize in economics has been awarded for some of these developments.

centage of blacks in a state's population. In quantitative terms, a 1 per cent increase in the proportion of blacks is associated on average with a reduction of $7 in the monthly amount. In other words, the higher the percentage of blacks in the population, the less willing the non-black majority appears to be to pay for welfare. But even this result depends to a large extent on the behaviour shown in just five states in the deep south (Georgia, South Carolina, Louisiana, Mississippi and Alabama), where both AFDC payments are very low and there is the largest percentage of blacks in the population in the whole of the US.

So we have evidence suggesting the existence of mild preferences on racial grounds by individuals, yet we observe marked racial segregation at the level of the system as a whole. This seeming paradox is entirely typical of many social and economic issues. It remains a paradox, however, only if we persist in thinking that we can automatically deduce the properties of the system as a whole from a knowledge of the behaviour of its component parts. This makes sense only if the economy and society are like giant machines, and all we need to do to predict and control them is assemble enough data and information and analyze it sufficiently cleverly. But if the individuals interact with each other, if their behaviour can be altered by the actions of others, we are in a type of system described at the end of the previous chapter. No matter how much information we gather, no matter how carefully we analyze it, a strategy of predict, plan and control will in general fail.

In the above evidence we have a particular example of such failure. Whether on racial grounds in America or class grounds in Britain, social segregation is both strong and persistent. Since Engels wrote about Manchester in 1844, the efforts of social reformers have not altered the basic divisions of the city. And the efforts since the Civil War to achieve complete integration in American society have not succeeded in breaking down all the divisions.

There is some degree of rationality in people in different income groups choosing to live in different neighbourhoods. Crime, for example, tends to be higher in areas where there is a high percentage of poor people. Indeed, this is the key reason why I myself chose in the mid-1980s to move from inner London to the leafy suburbs. Rioting had taken place in a large municipal housing estate some miles from where I lived, and property had been burnt down. From my window, I could see an estate built to the exact identical design. I felt, convenient though it was to be near the centre of London and my then place of employment, it was time to move.*

Further, there is evidence in the US that individuals do to some extent have a set of tastes and preferences in social matters that favour people of their own race. But these effects are weak. Individual preferences are only very mildly weighted towards members of their own race.

In contrast, we observe strong levels of social segregation over long periods of time. Economic theory suggests that such high and persistent segmentation requires powerful causal factors, rather than the rather modest ones which we appear to observe. But, surely, so too does common sense. The dramatic divide which we observe could presumably only arise from marked preferences by individuals. However, as so often, common sense is not necessarily a good guide to explaining social and economic phenomena. We can observe strong segmentation at the aggregate level, not because individual preferences strongly favour such an outcome, but because people have only a *weak* preference for it. Marked segregation can arise even when individuals are only mildly in favour of it.

Geographical segregation in itself, whether on class, racial or religious grounds, may, in fact, not be a problem. Indeed, it may well be inevitable so long as people have some choice over where they live. High degrees of segregation across a town or city can

---

* Cynics, of course, might note that, despite the criticisms which I make, I had after all received a comprehensive training in standard economic theory and so I was acting as a rational agent. But this merely shows that incentives matter rather than that agents respond to them in an optimal way. The area where I used to live has not in the meantime been put to the torch, so I could after all have continued to enjoy the benefits of geographical proximity to the centre.

come about from very mild preferences at the individual level. In other words, individual citizens may have a perfectly understandable wish to live amongst people similar to themselves, and this preference need not be very strong for the end result to be distinct segregation.

This original insight into the question of geographical segregation on racial lines was provided in a series of brilliant papers by the American Thomas Schelling thirty years ago. Like any theoretical model, Schelling's work abstracts from the fine details of the problem and concentrates on the key factors.

We can usefully think of theoretical models as maps which help us navigate around an unfamiliar area. The most accurate map which it is possible to construct would be of no practical use whatsoever, for it would be an exact replica, on exactly the same scale, of the place where we were. Good maps pull out the most important features and throw away a huge amount of much less valuable information. Of course, maps can be bad as well as good – witness the attempts by medieval Europe to produce a map of the world. In the same way, a bad theory, no matter how impressive it may seem in principle, does little or nothing to help us understand a problem. But good theories, just like good maps, are invaluable, no matter how simplified they may appear.

Schelling imagined a large chessboard, not of the usual eight by eight dimension, but much bigger. In his model, or game as we might think of it, there are equal numbers of two types of pieces which populate the board – 'agents' in the jargon of both economics and complexity theory. The types used can represent different social classes, racial groups or whatever. Initially, these are placed at random across the board, with a small number of squares left vacant. This is all that is needed to set the game up. A rule is now needed to describe how the game progresses, namely the rule agents use to decide whether or not to move. It is very simple: an agent decides to move from its square to a vacant one if less than a specified percentage of its neighbours are of the same type as itself.

The game progresses in a series of steps, with an agent being chosen each time to decide whether or not to move. If a neighbour is defined as one of the eight squares that surround any particular square, then the rule for moving is as follows: if four or more out of these eight squares are occupied by agents of the same type, an agent will not move. If there are less than four, the agent moves to a vacant square elsewhere. So an agent is content with its neighbourhood if, adding itself to its eight neighbours, five out of nine pieces are of the same type as itself. In other words, it is content if as many as four out of the nine are different.

In this version of the game, agents exhibit a high degree of tolerance for those of a different kind. They only move if they find themselves in a minority in their area. Yet across the chessboard as a whole, as the game progresses, the two types of pieces divide themselves into sharply segregated groups. From an initial configuration in which they are scattered around haphazardly, a very distinct pattern emerges very rapidly.

We can readily programme these rules into a computer to see what happens when we play the game. Schelling's achievement, it should be said, was all the more remarkable because he did it before the advent of modern personal computers and, indeed, obtained his key insights from calculations which he initially carried out by hand.

Figure 4.1 shows the initial random scattering of agents in Schelling's game, or model. There are an equal number of red and blue agents (shown in different tints, with the darker tint representing red and the lighter tint representing blue), allocated at random to different locations. We now set the game in motion and run it for long enough to allow each individual agent to have just two choices of whether or not to move.* The precise outcome will vary each time the game is played, but even with only two opportunities to move, the agents will separate themselves very distinctly into

---

* More precisely, the *average* number of choices across the agents is two. Because they are drawn at random, a few will not be asked to choose, as it were. More will have just one choice, lots will have two choices, some will have three and a very few may have four or five. But the average across them all is two.

FIGURE 4.1 Initial locations: the agents are allocated to squares at random. The two different types of agents are shaded (darker tint representing red; lighter tint, blue), and white squares indicate empty properties. 'Types' can refer to different classes, religious groups, racial groups or whatever.

neighbourhoods surrounded by their own type. Figure 4.2 is an entirely typical outcome of this game.

At the level of individual tastes, we observe only weak evidence of preference for members of one's own race. At the level of the system as a whole – a city, a country or whatever – we observe marked segregation. Schelling's model is a superb map, distilling the key features and enabling us to understand this apparent paradox. The model does not presume to tell us about the entire workings of the social and economic world, but focuses on the task at hand, namely, to explain why weak individual preferences are consistent with strong and persistent segregation.

The rules of the Schelling game appear to be – and indeed are –

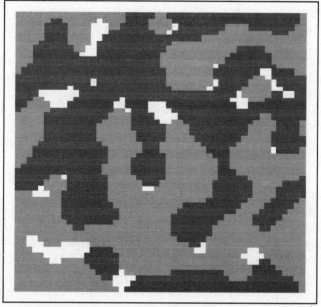

FIGURE 4.2 Typical pattern of locations after just two moves on average by each agent. Agents are happy as long as they are not in a minority in their neighbourhood. (Darker tint represents red; lighter tint, blue)

simple, but they give rise to complex, and unanticipated, behaviour of the system. This key, realistic feature of the game arises because the decisions of any one individual can impact in unexpected and unanticipated ways upon the decisions of others. A group of individuals is sitting perfectly happily in a neighbourhood. Unexpectedly, an agent arrives to fill an empty space. The newcomer may tip the balance, so that agents who were previously content now decide to move. In turn, their own moves may disrupt settled neighbourhoods elsewhere. And so the effects percolate through the system.

No single individual has intended, or even necessarily desires, this overall outcome. They are only concerned about what happens in their own local area. There, they are content to integrate, but the local interactions between them produce global segregation.

Importantly, we cannot predict the precise outcome of this particular game. We know that qualitatively it will lead to segregation, but the exact distribution varies in an unpredictable way from one solution of the model to another. Even if we started each time from exactly the same locations as in Figure 4.1 and solved the model many times, the corresponding Figure 4.2 illustrating the outcome would be different each time.

Each agent in the game acts according to a straightforward rule of behaviour. When we set in motion a solution to the game, we possess all the information that exists about it. We know exactly how each agent behaves. At any stage of the game, we know exactly what has happened. We cannot assemble teams of experts and committees to go out and investigate to find yet more data and examine yet more theories as to why we observe segregation. We know all there is to know. Yet, even so, we cannot predict the exact outcome of any particular solution of the game. And since we cannot predict it, we cannot control it. Plan, predict and control fails as a strategy, even though we have full and complete information.

Of course, the world is more complicated than that described in the model. In reality, for example, not everyone has the same ability to move and people are not scattered initially at random across a city. But this does not undermine Schelling's central insight: marked segregation can arise from only rather mild individual preferences. The 'unconscious tacit agreement' of Engels does not have to be strong at the level of the individual for the system of which the individuals are part to exhibit very marked levels of segmentation.

As it happens, within the confines of the theoretical model, Schelling's mild assumptions about individual preferences can be made considerably more extreme, and the overall outcome of the workings of the model are not affected. Romans Parcs of the London School of Economics and Nicolaas Vriend of the University of London have shown recently that neighbourhood segregation in the model will still arise even if individuals have strict preferences for perfect integration. The exact details need not concern us, for the paper of Parcs and Vriend runs to over fifty pages of rather dense mathematics, but it is an intriguing and, even to its many devotees, a rather startling extension of the model.

We can immediately think of additional ways in which the Schelling model leaves out some aspects of reality. After all, whether on class, racial or any other grounds, we rarely see in western societies the very sharp delineations between agents of different types of Figure 4.2. Some degree of mixing takes place. But, nevertheless, the central validity of the approach is clear. Despite all the attempts of social reformers to achieve greater integration, despite the countless policy initiatives over the years, segmentation persists. And it does so in the face of what in reality appear to be only mild individual preferences in favour of it.

In the previous chapter, we saw the key reasons why policy-makers cannot control the degree of inequality that emerges in a society. They operate in a very complicated world, in which decisions are enmeshed in inescapable uncertainty regarding their consequences. We saw how inequality evolves both at a global and national level, seemingly indifferent to the well-intentioned efforts of policy-makers to promote greater equality. Sometimes it improves, sometimes it deteriorates. Policies fail. Again, it is the sheer dimension of the problem, the fact that so many influences impinge upon it, which makes it difficult to comprehend and control.

Here we see, in a somewhat different guise, exactly the same force at work. Attempts to promote social integration, whether along class or racial lines, have largely failed, despite the fact that most individuals do not seem to feel strongly about the issues. They are happy to be integrated but, by an 'unconscious tacit agreement', this does not happen.

The Schelling model may seem simple, but it is a hugely complex system, whose outcomes depend upon the interactions between its component agents. No single agent in the model appears able to anticipate what might happen to the composition of its local neighbourhood, because ultimately the outcome depends upon the current and future decisions of every other single agent in the model.

Yet we can go much further than just invoking the dimension of the problem in order to appreciate the inherent unpredictability of the system. We have perfect and complete information about how individual agents behave. We know all there is to know in this particular context, yet we are still quite unable to predict the precise distribution of locations that will emerge in any particular solution of the model. Uncertainty reigns.

But is there perhaps a way of grappling with this malignant element, of bringing it under control? So far we have looked at firms, the distribution of income and now segmentation. In each case, we have identified an inherent element of unpredictability in the outcomes which appears difficult to overcome. In each case, this unpredictability, this uncertainty about the future, is due to the interactions between the players. The decisions of one agent can influence the well-being of others. It is these interactions that create the vast complexity which we observe.

The intellectual construct of game theory is designed exactly to illuminate circumstances in which individual agents, whether people, firms or governments, are pitted against each other, and in which the decision of one affects the outcomes for others. It will soon be time to turn our attention to game theory. But, first, what does economics have to say about uncertainty and its inseparable partner, failure? More precisely, what does the sort of economics that dominates the thinking of economic regulators and policy-makers around the world have to say?

5   Playing by the Rules

The problems that uncertainty creates for economics have been recognized for over a hundred years. In Chapter 1 we met the great English economist Alfred Marshall. He played a major role in developing economic theory in a systematic, scientific way. As already mentioned, his *Principles of Economics* exercised great influence in the opening decades of the twentieth century, and much of his work remains at the core of what is taught today in economics courses around the world.

Perhaps Marshall's greatest interest was in how markets work. The answer may seem obvious. Markets, after all, have characterized a great deal of human economic activity ever since we moved from being peripatetic hunter-gatherers to settled food-growers many millennia ago. The farmer brings his grain to market in the town or village, and the producers who turn it into bread offer to buy it. It seems a simple matter of supply and demand. In a year of poor harvest, when supply is low, the price will be high. And in years of abundance, it will be low.

But how does this market work? How is the price at which the sale takes place actually decided? Like many questions in science, the ones which seem easy to ask are often the hardest to answer, and it cannot be said that a completely satisfactory answer exists even today. Vernon Smith, Nobel prize-winner for economics in 2002, noted as much in his acceptance speech, when he stated perfectly frankly, 'We do not understand why markets work as they do.'

One hundred years earlier, Marshall realized that the answers to these questions, and in particular how the price is decided, were rather tricky. Not far from my house in the pleasant English countryside in Wiltshire is a superb estate. Rich with historical and cultural associations, this grand mansion nestles in seclusion in a sheltered valley. Servants' houses, stables and extensive game-shooting rights are associated with it. A few years ago, this unique property was offered for sale. Great secrecy surrounded the process, but it was eventually acquired by the singer Madonna for a sum rumoured locally to be some £9 million. How was this price decided? Very few people could afford the annual running costs of the estate, which must be many hundreds of thousands of pounds, let alone the purchase price. Of those fortunate enough to be in this position, most would be content with their modest apartments on Park Avenue, their Mayfair flats, their little villas in the south of France and the Hamptons.

In short, there was only a small number of potential buyers, and no matter what price was offered, the supply could not be increased, for the product being sold was virtually unique. But how was the price decided? It is hardly likely that the prospective purchasers were summoned to a meeting and made to bid against each other. Nothing so vulgar could have been contemplated. Yet, in essence, this must have been what happened. No matter how politely it was conducted, the potential buyers were involved in an auction against each other, the outcome of which, by its very nature, was uncertain. All those with a serious interest in the property were by definition stupendously rich. Yet why was the final price £9 million and not £8 or £10 million, or even £12 million? The only answer we can give is to say, quite simply, 'Because it wasn't.' It could very well have been an alternative figure, but the alleged £9 million was sufficient to deter other bidders and to secure the property for Madonna.

Marshall realized that this indeterminacy was present, albeit to varying degrees, in every market. Even with the Wiltshire country estate, with a very small number of potential buyers and a near-unique product for sale, supply and demand were important in determining the price at which the sale was concluded. But if there had been a different set of rich people bidding for the property, each equally wealthy as those who took part in the actual auction, the outcome could have been different, even if the number of potential buyers had been the same.

Marshall grasped that this might seriously limit the extent to which the results of economic theory could be generalized across markets. Yes, price was determined by the interaction of supply and demand, but the process by which this interaction took place seemed to mean that the outcome was uncertain. This in turn means that we cannot necessarily presume, for example, that an increase in the number of potential buyers – an increase in demand – would lead to a higher price in the market. It would all depend upon the process of the auction.

His contemporary, Francis Edgeworth, believed that such uncertainty characterized many real-world situations. Obviously, there would be limits to this uncertainty, to this indeterminacy. Madonna's fine shooting estate would always command a higher price than a suburban semi-detached house, but uncertainty of outcome would still exist.

The debate between Marshall and Edgeworth took place at the very time at which unprecedentedly large companies were springing into life. Substantial firms had existed previously but, as we saw in Chapter 1, towards the end of the nineteenth-century firms operating on a hitherto unimagined scale came into being. Many markets were becoming dominated to an increasing extent by a small number of very large firms, making reality seem much more like the situations described by Edgeworth in his theoretical work. He believed that, in such circumstances, there was inherent uncertainty about the outcome of the interplay between supply and demand. He developed a useful tool, the Edgeworth box, for thinking about it. In any given situation, there would be a range of potential outcomes for price, and his box was essentially a way of thinking about how the outcome might actually be decided. But the state of scientific knowledge at the time meant that he could take it no further.

Marshall's response was to try to assume away the problem. In this, he was tremendously successful. Marshallian diagrams of supply and demand, each with a uniquely determined price, have dominated economics textbooks ever since. In his world, any indeterminacy was erased out of the picture. It seemed that we *could* say after all that an increase in demand will lead to an increase in price and an increase in supply to a fall.* Economic textbooks often illustrate this with an example of a homely agricultural market such as wheat. In medieval times a bad harvest would mean a relatively small amount of wheat being brought into the city by farmers. Given the demand amongst the hungry populace, the price would have to be high to bring supply and demand into balance. In a good year for the harvest, the price would be low.

The textbooks encourage us to believe that we can readily substitute the word 'wheat' with, say, 'pick-up trucks' or even 'semiconductors'. In other words, we are invited to presume that the simple supply and demand diagram can be applied across the economy, in a wide range of markets. However, even in the context of a simple agricultural market, matters are not completely straightforward. For example, the amount which farmers are willing and able to supply today might not depend exclusively on today's price. With a poor harvest, the price is high, so they each might decide individually to plant more wheat for next year's crop. Given reasonable weather, next year's total supply will be high, and so the price will be low, which encourages producers to supply less the next year, and so on. In other words, the supply will depend upon the price in the previous year as well as this year's one.

In such circumstances, it is quite possible that the price will never be able to fulfil its function of bringing supply and demand into balance but will cycle around the equilibrium point. The potential for this to happen was realized many years ago and even used to feature in economics textbooks, where it was more usually illustrated with the example of the market for hogs rather than

---

* Actually, following the Sonnenschein-Mantel-Debreu theorems on market-demand functions of the mid-1970s, we now know, finally, that even in a completely Marshallian world we cannot presume *a priori* that these statements are true. Sadly, very few economics students are ever taught these difficult but fundamental results, established by the high priests of free-market theory. But the exposition of these theorems takes us far beyond the immediate purpose of this chapter.

wheat. The whole concept, however, appears to have been quite ruthlessly purged from most modern textbooks. Who is to say whether this is out of fear that students might no longer be able to grasp the arguments about the conditions under which equilibrium will not be found or out of the fear that they might feel that this undermines basic economic theory?

A more subtle problem arises from the fact that the supply and demand curves, so easily drawn in a theoretical diagram, are not in practice continuous. With the stroke of a pen, we can sketch on a piece of paper a demand curve that slopes downwards. This shows that the higher is the price, measured on the left-hand axis, the lower will be demand, measured on the bottom axis. The curve we usually draw is a continuous one, with no breaks, meaning that we presume to know the demand at *any* level of price within the range drawn by the curve.

In reality, however, prices do not take just any value. I cannot recall, for example, the last time I paid £π for a pint of beer or bought a CD for £e. Most people are familiar with π, though fewer ever encounter e, but both are extremely important in mathematics. Both are what is known in maths as irrational numbers. When expressed in decimals, the numbers never come to an end. We can approximate them, so that π for example is about 3.14 and e is about 2.72, but their true values go on for an infinite number of decimal places. Hence, there is a very good practical reason why I cannot remember buying things at these prices: it is literally impossible to do so. I might be able to buy goods for £3.14 or for £2.72, but not for £π or £e. This means that the demand curve I draw on a piece of paper as a continuous line can never exist. In reality it is not continuous but instead, in the language of maths, discrete. In other words, it is full of gaps, where numbers like π and e are. And, without going into the maths, *most* numbers are like these rather than like 3.14 and 2.72. So the curve is not a smooth one at all but looks a bit more like a sawtooth.

This may seem esoteric, but we can arrive at the same conclusion by purely empirical methods. We have only to look around a supermarket or a wine bar or a car showroom to see what types of prices actually exist in reality. No matter how finely graded the sawtooth might be in theory, it exists quite clearly in practice. We might see that, in a particular store on a particular day, a product is priced at £3.10. In another store, it might be £3.15 or £3.20. These days, it is unusual to see something at this sort of price with a label saying £3.14 or £3.16. Prices tend to move in jumps of 5 or even 10p. And we hardly ever see products priced at, say, £3.141. Indeed, the Dutch government has instructed retailers from 1 September 2004 to round all prices to the nearest 5 cents. There is a one-cent coin in the Euro currency, but the Dutch are effectively abolishing it and making it impossible to set individual prices in units other than of 5 cents.

More generally, retailers everywhere talk about 'price points' – standardized prices for particular types of goods. An example is £6.99 for a standard paperback book or £13.99 for a CD. The next price point may be quite a distance away – £19.99 for a specialized CD, for example, or £3.99 for a discount paperback out of copyright. Note the prevalence of 99p!

My own interest in the concept of continuity in economics, I should say, goes back to the years around 1970, when I was taught the subject in the then fervently anti-free-market faculty at Cambridge. High economic theory was our daily diet,* and part of it was the much more precise mathematical formulation of Marshall's supply and demand approach, which had been developed during the twentieth century. This clarified, along with much else, that we need the assumption of continuity† of demand and supply curves in order to prove that in free-market theory the price mechanism guarantees that supply and demand *can* balance. Without this assumption, the proof is not possible.

Criticisms of the conventional economic account of how markets work must not, of course, be confused with criticism of markets as such. Markets clearly work in practice, and work in

---

* We sat in the student café and sang, to the tune of Simon and Garfunkel's 'El Condor Pasa', 'I'd rather be continuous than discrete …'

† The actual continuity assumption needed is slightly less restrictive than this but does not get round the 'gap' problem.

general to the benefit of humanity. But, a century after Marshall, to quote Nobel laureate Vernon Smith again, 'We do not understand why markets work as they do.'

A hundred years ago the problems associated with what at first sight seems the simplest and most basic of all economic concepts were still in the future. Marshall appeared to be able to get over the challenge of indeterminacy laid down by Edgeworth, and his approach prevailed. His great professional prestige, occupying the chair of economics at Britain's leading scientific university at a time when the British Empire dominated the world, undoubtedly helped. So, too, did the fact that Marshall made economics seem more scientific. Economists appeared to be able, like natural scientists, to make statements that were generally true. In the natural sciences, for example, the theory of gravity is just as valid in Britain as it is on the opposite side of the globe in Australia. It seemed that economists could say that an increase in the number of buyers would lead to an increase in price, regardless of whether the market was organized in Britain or Australia.

The approach adopted by Marshall was to assume that there were so many buyers and sellers in a market that the addition or subtraction of any single one would make no difference to the outcome. No individual producer or purchaser could influence the price at which demand and supply were made equal. It is this simplification which led to the triumph of Marshall's approach.

The consequences of this seemingly arid intellectual debate have been profound. Not merely academic economics but also practical policies have been strongly influenced by it. The theoretical world created by Marshall is so special in economics that it has its very own name: perfect competition. It is yet another example of the terrifying ability of economists to brand their central concepts so effectively. Competition is generally believed to be a good thing, and if we can describe a theoretical world in which it is perfect, what else can we ask? It must surely be the aim of policy-makers to ensure that the real world is made in this very image, to be as close as possible to the nirvana of perfect competition. And, indeed, this is what economic regulators around the world usually try to do.

However, 'perfect competition' is simply a phrase, no more and no less. The words themselves give the theory no special status. Through the generations many economics students have abandoned the discipline on the reasonable grounds that the real world simply does not appear to be like this at all. Like any theory, it simplifies reality.

Two such simplifications in Marshall's world of perfect competition are particularly relevant to the problem of uncertainty. First, in this world, the product produced in any given market must be identical across all producers; not just very similar, but absolutely identical. In some markets, this is true. Pure gold bars of the same weight, for example, are identical in every meaningful respect. Once we move away from pure commodities, however, products differ. Consumer markets proliferate with different brands, different pack sizes, all subtly different from each other. One gold bar is just the same as another, but this is just not true with Weetabix and apple cinnamon Cheerios, for example. Both are breakfast cereals first and foremost, and to a visitor from Mars, say, they would appear to be virtually identical. But both have their separate sets of devotees to whom the differences are a matter of life or death (or at least, the cause of tears and tantrums at breakfast when the cupboard contains one but not the other).

The second simplification is a little more subtle. The assumption that no individual producer or purchaser can influence the price at which demand and supply are made equal may or may not be true in reality but, in a theoretical model, provided that we assume there is a very large number of both buyers and sellers, it seems quite reasonable. Yet this is where a fundamental problem lies. We can discuss it in the context of the application of modern mathematics and the theory of infinitesimals. More graphically and more simply, we can again quote Vernon Smith, 2002 economic Nobel Laureate: 'As a theory, [this] is a non-starter: who makes price if all agents take price as given?' In other words, if by assumption no single buyer or seller can influence the price, how is the price ever decided? So, even on its own terms, the Marshallian world contains an internal contradiction.*

The intellectual tool of game theory, developed during the second half of the twentieth century, appears to offer a way round these fundamental problems of uncertainty and indeterminacy. It seems to offer the possibility of being able to handle uncertainty in a rational way. As a result, it has been seized upon by economists. Academic conferences and journals are replete with papers on game theory, and reputations and careers are built on the ability to manipulate the abstruse mathematics that are frequently involved.

In the auction, if we may call it that, which took place during the purchase of Madonna's English country house, the best price for the winner to pay would be just a single British pound more than the second-placed person. However, each of the potential purchasers were acting under uncertainty. They did not know in advance the price at which each of the others would drop out and the prize would be theirs. Game theory offers, in principle, a way of devising rational strategies for bidding in such circumstances.

It can be used equally by sellers as well as buyers. For example, when the British government carried out an auction for 3G mobile (cell) phone spectrums in 2000, the structure of the auction was devised by game theorists. Some of them believed that their design would lead to an optimal outcome for all concerned, for both the government and the competing firms. Certainly, from the point of view of the British government, trying to maximize the takings, it was a spectacular success, raising over £20 billion for the UK Treasury, far more than anyone had imagined in advance. But despite the fact that each of the bidders themselves had their own team of game theorists advising them how to behave, in a wider context its success was more questionable. The high-technology firms who took part paid huge amounts of money, far more than

---

* This is not to deny Marshall's greatness as an economist. His words, as opposed to his diagrams and his maths, contain huge numbers of insights into how the world works which are still relevant today. In the particular context of this book, he believed that economics had a great deal to learn from biology.

they had anticipated, simply to secure the rights to operate, money which could otherwise have been used for investment or research and development.

When supermarket managers, selling goods to the public, decide the price at which a particular brand of baked beans, say, will be offered, they are playing at least two elaborate games, one with their customers and the other with the managers of rival super-market chains. As we have seen, in reality the economic world looks much more like that described by Edgeworth than Marshall's world of perfect competition. Buyers and sellers by their individual actions can influence the price at which products and services are sold.

The intellectual origins of game theory go back some sixty years, to a man who was possibly the greatest polymath of the entire twentieth century. John von Neumann was born in Budapest in 1903. He first moved to the US in 1930, finally settling there permanently in 1933 when the Nazis came to power in Germany. Even by then, he had made strikingly original contributions to a wide range of topics. Not merely the contents but many of the titles of his scientific papers are incomprehensible to most people. The spectral theory of Hilbert space, ergodic theory, rings of operators, measures on topological groups and almost periodic functions were amongst his interests in mathematics. He had also worked on logic and quantum mechanics.

Von Neumann's intellectual energies blossomed even more in America. He was actively involved in the Manhattan project, the construction of the atomic bomb in the 1940s. He made a great contribution to the development of the computer. He anticipated by decades modern interest in the mathematics of chaos theory. In many ways the archetypal mad scientist, he was involved in multifarious ways with the US military and the CIA. These activities led him to become, allegedly, the role model for Peter Sellers's portrayal of the central character in the classic film *Dr Strangelove: or, How I Learned to Stop Worrying and Love the Bomb*.

Almost in passing, von Neumann took an interest in some very difficult problems in economics. In a seminar at Princeton in 1932, for example, he placed free-market economic theory for the first time onto a modern mathematical basis, illustrating in fact the crucial importance of the assumption of continuity of demand and supply curves, which we discussed above. This represented a major step forward in understanding not the strengths but the limitations of such theory. By analyzing it rigorously on its own terms, he was able to establish far more precisely than had ever been done before the conditions under which pure free-market theory could be said to be a reasonable description of reality.

However, his best-known work amongst economists – it is implausible to claim that any of von Neumann's work was on the bestseller list – is a collaboration with Oskar Morgenstern published in 1944. *The Theory of Games and Economic Behavior* was embraced enthusiastically by the economics profession. The influence of the concept of game theory on economics has grown dramatically over time, with the years since 1980 in particular seeing an explosion of interest in the topic. For game theory appears to offer economics a rational, calculable way of dealing with uncertainty.

Uncertainty surrounds most economic decisions. It pervades areas of social concern and confuses policy-makers in both the public and private sectors. It is the principal cause of failure. Any method which holds out the promise of coping with the problem is bound to seem attractive. Game theory appears to fit this bill.

Outside the world of business and economics, game theory might also help to illuminate military strategy. In the Cold War of the second half of the twentieth century, the US and the Soviet Union were playing a multi-layered game. Both powers were armed to the teeth with nuclear weapons. Curtis LeMay, head of the American air force, had stated that the Soviet Union could be reduced to cinders with six thousand nuclear warheads. Yet he wanted the money for ten thousand. At a formal hearing, an elected representative had the temerity to remind LeMay of his statement. His reply was curt: 'Senator, I want to see those cinders dance!' And this was indeed the point. Both sides had the power to destroy completely the other. Making a gigantic leap and leaving aside moral considerations, might not the optimal strategy have been to do precisely that in a massive first strike, rather than living with a perpetually uncertain stand-off?

The closest the world came to nuclear war in that period was probably the crisis of 1962, when Khrushchev planned to install missiles in the Soviet client state of Cuba. He and President Kennedy played a tense game, which as we know ended in a peaceful conclusion. But the veil of uncertainty enveloped each stage of this nerve-racking process. If we do this, will he do that or will he do something else? That was the question facing both sides. Both actors could be assumed to be rational in the sense of preferring to avoid a nuclear exchange if at all possible, so a game-theoretic approach appears to offer a way of coping with the uncertainty and deciding strategy.

These military examples are not given by accident. The 1944 book by Morgenstern and von Neumann on game theory was an exciting new intellectual development and, as is so often the case with powerful new ideas, the US Department of Defense harboured some of the first people able to see its potential. A specific application was perceived almost immediately to be the stand-off with the Soviet Union and the strategy on nuclear war. The RAND Corporation think-tank at Stanford gathered a group of distinguished scholars to develop the game-theoretic approach, both in the specific context of nuclear strategy and more generally.*

Two of these, Merrill Flood and Melvin Dresher, invented the Prisoner's Dilemma game in January 1950. The game is rather straightforward to describe, but fiendishly difficult to play. Even by the end of the 1970s, it had given rise to over a thousand articles by academics from a whole range of disciplines, such as economics, maths and psychology. The torrent has since become a veritable tidal wave.

At first sight, it seems wholly implausible that such a simple game could generate such a level of academic interest. Imagine

---

* Philip Mirowski's book *Machine Dreams* (Cambridge University Press, 2002), an original and immensely scholarly book, gives an excellent and detailed (two words which don't always go together!) account of this fascinating period.

that you and your partner devise a plan for robbing a bank. You steal a high-powered getaway car and carry out the crime. Unfortunately, you are caught by the police. They lack conclusive proof that you have committed the crime, but strongly suspect you both. The action, as in most such cameos, takes place in the US, where the authorities have the power to make bargains with criminals in return for confessions. You are each held in a separate cell, and a deal is put to you. As far as your partner is concerned, you are completely in the dark. You are wholly uncertain about anything that is being suggested to him or her, but you are compelled, for the purposes of this story, to decide on the spot what you will do.

The offer from the prosecutor is this: you can either confess or remain silent. If you confess and your partner remains silent, you get off completely free, and your testimony will ensure that your accomplice gets a very long jail sentence. The same applies in reverse. If you keep quiet and your partner confesses, it is you who serve the time. If you both confess, the legal authorities will ensure you get a fairly substantial sentence, but one which is much shorter than you would otherwise get. Finally, if you both remain silent, the prosecutor will have to settle for a very short sentence by convicting you of merely stealing a car.

In the jargon of the game, keeping silent is referred to as a strategy of co-operation. If you both keep silent, you get the best deal for both of you. The outcome is the same as it would be if you could both liaise and agree to co-operate with each other. Confessing is known as a strategy of defection, for by doing this you are potentially defecting on your partner and making things much worse for him or her. So keeping silent signifies co-operation, and confessing signifies defection.

In a nutshell, these are the rules of the Prisoner's Dilemma. Sitting in your cell, completely uncertain about what your partner is doing, what strategy should you choose?

In fact, in this very simplest version of the game, there is a best strategy – an optimal one, as economists like to say – for you both to follow. This is for you both to confess. You do, admittedly, incur a jail sentence if you follow this strategy of defection. So why is this the best strategy to follow? The key is that you are operating under uncertainty; you do not know what your partner in crime will do.

Your very best policy is to keep quiet, to co-operate, but only provided that your partner does the same. If you do this and he defects, you personally face a very long jail sentence. If, on the other hand, you defect and he co-operates, you get off completely free. And if you both defect, then you get a modest but not too onerous jail sentence. So, in the face of uncertainty, the best strategy for you both is to defect.

This sums up the dilemma faced by the criminals. Whatever the other does, each is better off confessing than remaining silent. But the outcome obtained when both confess is worse for each than the outcome they would have obtained had both remained silent. This is a rather unwelcome conclusion. There appears to be an inherent conflict between individual and group rationality. It is rational for the group – the criminals – to co-operate, but for each individual criminal it makes sense to defect.

This optimal strategy was worked out very quickly when the game was first invented. It seems to suggest that the best thing to do in a nuclear race is to defect and launch a first strike. One hopes that the director and writers of *Dr Strangelove* were familiar with this result, for this is exactly how the film ends. *Ars longa, vita brevis!* Art can indeed take its inspiration from anywhere, even from the abstruse maths of game theory.

However, the proof that this is the best strategy for individuals to follow depends upon a very important assumption. In the case of our prisoners, the game is only ever played once. It is wholly implausible, even by the standards of Hollywood, to imagine that the two players will find themselves at some point in the future in the identical situation. Yet in most real-world situations, even where the Prisoner's Dilemma might be thought to have a useful application, we simply do not know how long the game will last. Should two competing firms, for example, start a price war or should they each try to maintain their price levels? There are many practical problems in applying the Prisoner's Dilemma here, but the most important is that neither firm knows for sure how long the game will go on for.

Introducing uncertainty about when the game ends introduces a whole new dimension of uncertainty. So vast and complicated is this new dimension that even today the 'best' strategy to follow remains unknown. The scientific community has invested a great deal of effort in trying to discover the best strategy, but still we do not know.

Robert Axelrod, a political scientist at the University of Michigan, has been instrumental in stimulating the massive research programme into this game over the past twenty years or so. Initially, he conducted a computer tournament where people were invited to submit strategies for playing 200 games of Prisoner's Dilemma. Fourteen game theorists in disciplines such as economics and mathematics submitted entries. These strategies, and a totally random one, were paired with each other in a round-robin tournament. In other words, they all got to play each other. Some of these strategies were highly intricate, but the simplest of all strategies got the highest average score. This strategy, called Tit for Tat by its developer Anatol Rapoport, had only two rules: on the first move co-operate; on each succeeding move do what your opponent did the previous move. So if its opponent at any stage played co-operatively, Tit for Tat would do exactly the same, raising the possibility of the two strategies getting locked into a sequence of mutually beneficial co-operative moves. This is exactly what happened and was essentially why this strategy scored more highly than any other.

After publishing these results, Axelrod organized a second tournament. This was exactly the same as the first, except that the games ended at random, with an average number of 200 moves per game. This time there were sixty-two entries. In addition to mathematicians and economists, strategies developed by computer scientists, evolutionary biologists and physicists were also present. Rapoport again submitted Tit for Tat and again it won.

The outcome of the tournaments was exciting. The conflict between individual and group rationality appeared to have been resolved. A strategy of co-operation, with the sanction of defecting against players who initiate a defection themselves, seemed to be best.

Unfortunately, subsequent work shows that we cannot necessarily draw this conclusion. The degree of uncertainty in the Axelrod tournaments was low. Indeed, in the first tournament it was effectively non-existent. The strategy of each player remained fixed throughout, and each game ended after a known number of moves. The second tournament allowed just that bit more uncertainty, by permitting games to end at random, but the average number of moves across a sequence of games was fixed and known.

Much more uncertainty can be introduced very easily. For example, the length of playing time between any pair of strategies can be made genuinely random. In addition, the strategies of each player may not be fixed but be allowed to develop mutations over time. Further, completely new strategies may enter the tournament at random at any time.

These complications make the game more realistic, more true to life. Think, for example, of rival supermarket chains, each with outlets competing head-to-head in towns and cities across the country. Neither is compelled to follow the same strategy week in, week out. Indeed, they will almost certainly vary their strategies, with price promotions on different baskets of goods, advertising campaigns, introduction of a wider range of brands, or even entirely new features such as an in-store butcher or fishmonger, and so on. And there is always the risk that in some of the locations another competitor might open a new store, following its own particular strategy for success.

Once the level of uncertainty is raised, it becomes much harder to decide whether strategies based primarily on co-operation or defection are the better. Indeed, in the current state of scientific knowledge, we simply do not know the optimal strategy for the game.

All this suggests that the Prisoner's Dilemma, a very simple game to describe, leads to such complications in its play that in most situations it is of not much practical use. In certain highly simplified circumstances, knowledge of the game is invaluable. In the demotic language of the old Cockney gangs of the East End of London, criminals were honour-bound never to 'grass' to the police. However, so attractive was the policy of defection in this particular context for certain individuals that as a result most of the gang members ended up serving long prison sentences. But, this specialized application aside, the game can hardly be said to be of value to practical strategic decisions when even the many thousands of academic articles on the subject have failed to illuminate the best type of strategy.

An example of the latter was posted, as I was writing these very paragraphs, on the world econophysics website,* where physicists and mathematicians from around the world with an interest in economic and social problems post their papers. The title itself is sufficient to deter most would-be readers: 'Cooperative Equilibria in the Finite Iterated Prisoner's Dilemma'. The complexity of the analysis which this simple game has generated may be grasped from the summary – not the actual dense mathematical analysis, but the summary in words – of this particular paper, which is entirely typical of the genre. I do not propose to translate it into everyday English, for this task would fill many pages, so here it is as written:

> Nash equilibria are defined using uncorrelated behavioural or mixed joint probability distributions effectively assuming that players of bounded rationality must discard information to locate equilibria. We propose instead that rational players will use all the information available in correlated distributions to constrain payoff function topologies and gradients to generate novel 'constrained' equilibria, each one a backwards induction pathway optimizing payoffs in the constrained space. In the finite iterated prisoner's dilemma, we locate constrained equilibria maximizing payoffs via cooperation additional to the unconstrained (Nash) equilibrium maximizing payoffs via

* http://www.unifr.ch/econophysics, for those readers with an aptitude for heavy-duty maths.

defection. Our approach clarifies the usual assumptions hidden in backwards induction.

This has not deterred economists from seizing on the concept of game theory, believing that it enables them to tackle the uncertainty that is inherent in the real world and so to preserve the idea of 'rational' individual behaviour.

One concept above all is revered by the theorists who devote their lives to this topic: the so-called 'Nash equilibrium'. John Nash was a highly eccentric mathematician who received the Nobel prize for economics in 1994 for his invention of this idea. He has become much more famous than most economists through the popular film and book of his life, *A Beautiful Mind*, starring Russell Crowe as the tortured and increasingly bizarre central character.

Nash discovered his concept almost at the same time as the Prisoner's Dilemma was born. Von Neumann was characteristically scathing, dismissing the idea as being 'trivial, it's just a fixed-point theorem'. Now, even the simplest of fixed-point theorems involves some pretty hard maths, but as we know, almost everything was trivial to von Neumann. Translated into plain English, Nash's idea seems very sensible. A Nash equilibrium in a game is a set of strategies, one for each player, such that no player has the incentive to unilaterally change his or her action.

For economists, of course, the concept contains the magic word 'equilibrium'. The world may be fraught with uncertainty but, armed with Nash's equilibrium, the idea of rational, calculable and certain behaviour can seemingly be restored. Players, whether people, firms or governments, must surely act rationally and seek to find a strategy that means that they themselves are as well off as they can possibly be, given how everyone else in the game is behaving. An enormous number of academic papers on economic theory have appeared in which the Nash equilibrium plays a central role.

There is one slight problem with this marvellous concept: real people do not seem to know about it. More precisely, they frequently adopt strategies in a game context which most definitely do not correspond to those of the relevant Nash equilibrium.

Some games are, of course, very much easier than the Prisoner's Dilemma. Noughts and crosses, for example. The outcome of this game should always be a draw since most combinations of moves will lead to this conclusion. In technical terms, this means that there is not just one but several Nash equilibria; in other words, several strategies such that, once adopted, no player has an incentive to change from them. But noughts and crosses is so easy that even a chicken can find a Nash equilibrium. In 2003, a new casino in America had a marketing ploy to attract punters. They got to play a single game of noughts and crosses against a chicken. If they beat the chicken, they won $5,000. Grains of corn – or whatever it is chickens like – were scattered across a board, and the chicken made its mark by pecking in a square. In the first month of operation, only one human won the $5,000. In other words, a 'player' with a notoriously low level of intelligence – a chicken – is able to play this game almost as well as a player with much higher intelligence, namely a human. Clearly, the best strategy in this game must be exceptionally easy to find. In practice, a chicken can find it by choosing squares, not according to how best to play the game, but according to where its food happens to be scattered. Indeed, it *is* very hard to lose when playing this game, since in most situations most of the choices which are available will lead to a draw.

As we have seen with the Prisoner's Dilemma, once things get a bit more complicated, playing a game becomes very much harder. Game theory can certainly make a contribution to, say, poker, and anyone intending to play in tournaments had better be familiar with the major texts (a path-breaking one had the irresistible title of *How I Made Over $1,000,000 Playing Poker*). But theory can only take us so far.

The fundamental dilemma for all poker players is quite simply: has he got a hand or hasn't he? And as Amarillo Slim, one of the very first official world champions of the game, once put it: 'All trappers don't wear fur caps.' Translated, this pithy and effective Texan phrase means that someone may have an unbeatable hand, but will indulge in positively Shakespearean acting in order to trap the other players and lure them into the pot. Bluff and counter-bluff are fundamental features of the game in complex ways which are virtually impossible to characterize. At the top level, everyone knows the game-theory textbooks, but game theorists rarely if ever walk away with the million-dollar pots.

A clear example of how an apparently simple game proves hard to play in practice is given by *The Price is Right*, a very popular television game show in the US and many other countries. The rules are very straightforward and easy to remember. In other words, players have full knowledge of the rules. At all times, each player knows the state of the game. In addition, we can be sure that all those who actually get to play the game on television are devotees. They will have previously watched many episodes, shouting out advice or derision at the contestants from the comfort of their television rooms at home, and will have had every opportunity to consider good strategic moves.

Two academics, Rafael Tenorio and Timothy Cason, worked out analytical solutions for the Nash equilibrium strategy in *every* possible play in the game, a truly impressive intellectual feat. Their results were published in the *Economic Journal* in 2002. Even more interestingly, they went on to compare these with the outcomes of what actual players did in some 300 editions of the programme. (They experienced some difficulty in obtaining tapes of the show from the producers. Their reason for wanting them – that they were going to compare actual decisions in the game with the appropriate Nash equilibrium move – did at first sight stretch credibility. Eventually, however, they succeeded.) They discovered that, except where the Nash strategy is trivially obvious, as it is, for example, in noughts and crosses, most of the time most of the players did not find it. Sometimes, their actual strategies were far removed from the optimal Nash decision.

The Price is Right is not a difficult game. The dimension of the problem might not seem to be large *a priori*: the rules are clear; there is no uncertainty about the situation in which a decision has to be made; and each contestant is in possession of full information about it. Yet, in practice, people with every incentive to succeed usually failed to compute the Nash equilibrium.

The disjuncture between how people ought to behave according to game theory and how they actually do behave is not a modern discovery. As Philip Mirowski makes clear in his book *Machine Dreams*, experiments at RAND established this almost as soon as games such as the Prisoner's Dilemma had been invented over fifty years ago. Indeed, Merrill Flood, its inventor, soon abandoned work on game theory altogether for exactly this reason.

Two examples will suffice. In the first, Flood offered RAND secretaries a choice. One of them was given the option of either receiving a fixed sum of money ($10, say) or receiving a total of $15 provided that agreement could be reached with another secretary as to how this money was to be divided between them. One Nash solution is that the two split the marginal difference. In other words, they divide the extra $5 between them so that they get $12.50 and $2.50 respectively. Obstinately, in practice most secretaries were attracted not to the new idea of the Nash equilibrium but to the concept of fairness, as old as humanity itself. They divided the total amount exactly equally, $7.50 each.

The second is even more interesting. Flood carefully devised a pay-out system in the Prisoner's Dilemma in which the best option for both players was not the usual co-operative one; the Nash equilibrium was unequivocally for both players to defect. To play the game, he recruited distinguished RAND analysts John Williams and Armen Alchian, a mathematician and economist respectively. They were to play one hundred repetitions of the game. They each knew about von Neumann's work but not about the Nash equilibrium, which had only just been discovered. Both were asked to record their motivations and reactions in each round.

The Nash equilibrium strategy ought to have been played by completely rational individuals a hundred times. It might, of course, have taken a few plays for these high-powered academics to learn the strategy, but Alchian chose co-operation rather than the Nash strategy of defection sixty-eight times, and Williams no fewer than seventy-eight times. Their recorded comments are fascinating in themselves, and a single aspect will have to suffice us here. Williams, the mathematician, began by expecting both players to co-operate, whereas Alchian the economist expected defection. But as the game progressed, co-operation became the dominant choice of both players.

Nash was immediately told of these results, and his reaction is quoted at length by Mirowski. Many of the points are technical, but the most dramatic by far is the following: 'It is really striking how inefficient the players were in obtaining rewards. One would have thought them more rational.' In other words, his theory predicted a particular kind of behaviour. The players did not follow it and, clearly, the mistake lay with them and not the theory. Two very clever people, intimately familiar with game theory in general, had persistently chosen a non-Nash strategy. But the theory simply could not be wrong, because that is how rational people ought to behave!

Despite the fact that Nash was primarily a mathematician, it is perhaps fitting after all that he received the Nobel prize for economics. The attitude that the theory cannot be wrong and that it must be the people who are wrong is one which is entirely typical of a large number of economists. Unfortunately for economists, to paraphrase Bertolt Brecht, we cannot elect a new people.

It is not that game theory is entirely useless in social and economic contexts. In certain highly specific, structured contexts it can be a powerful method of analysis. One of the best books on the theory in my opinion is written by a game-theory enthusiast. David Kreps' *Clarendon Lectures* were given in 1990, but the problems he identifies for game theory have not really been dealt with in the subsequent decade or so since then. This is why it is such a good book. There are lots of technical details, but he is at pains to stress the serious limitations of the game-theoretic approach. It cannot inform us about an enormous range of practical situations. A key point he makes is that 'game-theoretic techniques require clear and distinct rules of the game'. Kreps believes this is so important that he puts it in italics.

For game-theoretic techniques to be of value, clear rules of the game are needed. But where do these rules come from? Do they fall from the sky like manna from Heaven? Of course not. It is one thing to use game theory in a highly organized context such as a governmental auction for cell-phone bandwidths. These settings are designed deliberately to minimize the level of uncertainty faced by participants. The rules are clear, and the criterion for success is unequivocal. Either your firm obtains a licence to operate, or it does not. In most practical contexts, however, reality itself is diffuse, fuzzy, shaded at the edges. It lacks the very clear and firm guidelines which game theory needs to be of any possible use. And even then, as we have seen, very simple games have not yielded much information on the best strategies to follow. Game theory helps in penetrating the veil of uncertainty, but it simply lifts one tiny part of the curtain, leaving most of the stage in darkness.

A key paradox begins to emerge from all this. Humans, whether acting as individuals or making collective decisions in companies or governments, behave with purpose. They take decisions with the aim of achieving specific, desired outcomes. Yet our view of the world which is emerging is one in which it is either very difficult or even impossible to predict the consequences of decisions in any meaningful sense. We may intend to achieve a particular outcome, but the complexity of the world, even in apparently simple situations, appears to be so great that it is not within our power to ordain the future. '*Ordina quest' amore, o tu che m'ami!*' is the prayer of the Italian medieval poet Jacopone da Todi, which Dante famously set at the head of his *Purgatorio* ('Set my love in order, o thou who lovest me!'). But lacking the divine intervention on which the poet calls, we have no means to guarantee that we will achieve the desired outcome. Given that as humans we are able to gather information, process it and take informed decisions on the basis of this, how can this be? It is this problem which we will now begin to consider.

6 A Game of Chess

Our understanding of the consequences of decisions we take today on our future well-being, on our fitness for survival, appears to be very limited. Companies want not merely to survive but to make profits and prosper. Yet, as we have seen, they often fail. Governments want to achieve all kinds of desirable social goals, yet they, too, often fail.

This confronts us with a paradox. Humans can take decisions with intent, acting with the purpose of achieving specific targets. As we noted in the Introduction, this ability to act with intent is sharply different from the process of biological evolution, which takes place at random. Yet both cases, whether human strategy or the evolution of species, are characterized by widespread failure. The human ability to act with purpose and intent seems not to imply in any way that the actual outcome will be the desired one. This raises a fundamental question: what are the factors which limit the effectiveness of purpose and intent, and which often lead to outcomes which are either unexpected or undesired?

The question of intent and purpose in human action is one of the oldest and most problematic that exists. Martyrs have quite literally gone to the stake over the matters to which the concept gives rise. In the religious conflicts of Europe in the sixteenth and seventeenth centuries, in which Protestantism broke with and replaced Catholicism in a number of areas, huge numbers were slaughtered and regions laid waste in conflicts which occurred over disputes of doctrine. The fire and the sword have been the instruments of avenging, conquering armies for millennia. The development of new and terrifyingly powerful instruments of war made the devastation even worse. Whole areas of central eastern Europe, for example, were annihilated during the course of the

Thirty Years War in the seventeenth century, in ways which set back their economic development by at least a hundred years.

At the time of the east Asian economic crisis in late 1997, I was in Jakarta, the sprawling and dynamic capital of Indonesia. After giving a lecture, I was being shown round the city by a group of very polite and helpful young men. We passed the Catholic cathedral. Then a Protestant church was pointed out. 'What', they asked, 'is the difference between the two?' In the oppressive tropical humidity, the energy was lacking to explain the subtle but crucial distinction between the corporeal and merely symbolic interpretation of the bread and wine at a particular point in the Communion service. Even as the phrases formed in my head, I somehow doubted that I would be believed, so I simply answered, 'It's the Pope.'

I ought to have engaged them in a philosophical debate about free will and predestination. It was this, above all, that lay at the heart of the European conflicts several centuries ago. It was a key reason for the flight to America of the Pilgrim Fathers in the *Mayflower*. From one perspective, God's elect are predestined from the beginning of time for salvation. This view led to many agonized and heated debates about how the elect were to be recognized during their time on Earth. Perhaps they were programmed, to use a modern phrase, to carry out good works. But, if so, could people consciously choose to behave well, and in so doing demonstrate their membership of the elect? It seemed unlikely, for this would contradict the very notion of predestination, of predetermined behaviour. In contrast, believers in free will held that individuals were entirely responsible for their own actions. They could choose how to behave, and it was these choices which would ultimately be judged.*

In a less abstract context, we need to consider the parallel between biological evolution and human social and economic behaviour as far as agency and intent are concerned. Assuming for

* Anglicans, of course, are free to believe either, both or neither of these doctrines. However, they ought to be, like Keith Richards of the Rolling Stones and now a firm Anglican, devotees of the service of Choral Evensong.

the moment I am not predestined to behave in this way, I can act with intent at this very moment. I can stop writing this chapter and watch television or go to a wine bar, or whatever. In the same way, an animal could decide to, say, stop searching for prey and have a rest instead. Even this seemingly innocuous statement is not without controversy. As Thomas Nagel noted in his classic 1974 article 'What Is It Like to Be a Bat?', 'Some extremists deny the existence [of conscious experience] in mammals other than man.'

Nagel's paper, published in *The Philosophical Review*, deliberately chose the example of a bat to discuss the question of consciousness. The bat is a mammal, a property it shares with humans. Nagel argues that if one travels too far down the phylogenetic tree, people gradually shed their faith that there is conscious experience there at all. So he selected bats rather than wasps or flounders. But bats are otherwise completely unlike humans. Most bats perceive the external world primarily by sonar, detecting the reflections, from objects within range, of their own rapid, subtly modulated, high-frequency shrieks. Their brains are designed to correlate the outgoing impulses with the subsequent echoes, and the information thus acquired enables bats to make precise discriminations of distance, size, shape, motion and texture comparable to those we make by vision. However, bat sonar, though clearly a form of perception, is not similar in its operation to any sense that we possess, and there is no reason to suppose that it is subjectively like anything we can experience or imagine. In Nagel's laconic phrase, 'This appears to create difficulties for the notion of what it is like to be a bat.' We simply do not know whether bats can behave with intent or, rather, what it means for a bat to believe it is acting with intent.

Many deep and unresolved issues surround the concepts of consciousness and intent, of free will and predestination. But, in the true spirit of English empiricism, we can move on by pushing them firmly to one side. In this, we are at one with economists, who certainly believe as a matter of faith in the purposeful and conscious nature of human decisions.

The relevant concept in an evolutionary context is in any event rather more straightforward. It is not a matter of whether a bat is capable of conscious thought and planning its actions. Rather, it is whether the animal has any control over its evolution, such as how its sonar develops or deteriorates. Clearly, the answer is 'no'.

In a social and economic context, there is a certain amount of fuzziness involved, but, again, the answer seems fairly clear. The ability of an economic agent, a firm or individual to prosper, its fitness to survive, can be affected by decisions which are consciously taken by that agent. It is not so much a question of my ability to decide to get up off the chair and switch on the television, but which house I buy, which pension scheme I decide to take, or the price at which I set my product or locate my office or factory.

We might usefully reflect on the chilling scene in *Macbeth*, when the eponymous hero is about to go and murder King Duncan. Night has fallen and Duncan has retired to bed. Macbeth stands, firm and resolute, emboldened by the encouraging words of his sinister wife. 'I go, and it is done,' he announces. Macbeth here acts with the intent to carry out a strategic decision: to kill the king. He believes this will act to his competitive advantage. In this particular context, urged on by his wife, he believes that the act will assist his own progress to the regal throne. Indeed, shortly afterwards, he does become king.

But matters are not necessarily so clear-cut. For dramatic purposes, of course, when Shakespeare has Macbeth pronounce his deadly intent, it is necessary for him to be successful in his course of action. The reality might have been rather more difficult. A medieval monarch arrives to stay at the castle of one of his most powerful barons, who has a plan to murder him. In practice, the king may post a secret guard in his bedchamber. He may, as a matter of routine prudence, have several rooms made up and make his final choice at random, so no one else knows exactly where he is sleeping.* Less deviously, Duncan may have suffered from

* A practice followed successfully for many years by the late J. V. Stalin.

insomnia, so, when Macbeth creeps into the room expecting a slumbering king, he is confronted instead by Duncan reading the Venerable Bede's *Ecclesiastical History of the English People* by candlelight.

Any of these rather mundane but practical possibilities would certainly have spoiled the play. None is implausible. The baron formed a strategy to kill the king and intended to carry it out. In practice, however, he would have faced very considerable uncertainty. Any of the above suppositions, from the deliberate planting of a guard by Duncan to the somewhat more slapstick idea of an insomniac monarch, could have frustrated Macbeth's plan. The sinister phrase 'I go, and it is done' makes for tense drama, but even over the course of the short walk then taken by Macbeth into the bedchamber, uncertainty of outcome existed.

Speculation aside, the denouement of the play itself reveals the pervasive existence of uncertainty. Macbeth killed Duncan, believing that this would advantage him, would enhance his own fitness for survival. He successfully carried out his self-appointed task. In the longer run, however, this led not to his enhanced survival but to his downfall and eventual death. Macbeth believed he was carrying out a strategic act that would work to his advantage, but this very act caused other agents in the play to alter their own decisions in ways in which the regicide and his evil wife had not considered. In short, the eventual impact on Macbeth's fitness for survival was the opposite of what he intended. The degree of uncertainty that he faced was too large for him to take proper account of it in his fateful decision.

Reflecting more generally on such matters, most of the world's great literature is based upon the uncertain and unexpected consequences of individual actions. The whole of Virgil's *Aeneid*, for example, would lose its point if Aeneas knew from the outset that he would fall in love with Dido and then betray and leave her. For example, in the great sixth book, Aeneas is granted spiritual initiation into his new realm of Italy and is empowered to be guided into the Underworld by the Sybil. There he encounters the shade of Dido, who denounced him when he abandoned her, vowing that 'When cold death has severed soul and body, everywhere my shade will haunt you.' Now, she has killed herself and is in Hades. Aeneas attempts to defend himself, but instead is dealt the most monumental snub in literature: '*Illa solo fixos oculos aversa tenebat*' ('She turned away, keeping her eyes fixed'). Far from even speaking to him, despite his immense efforts to locate her in Hades, Dido merely glares, resolutely turns her back and runs off to her husband in Hades. The dramatic effect would be entirely lost if Aeneas had anticipated this when he landed on the shores of Carthage in the opening book, many hundreds of lines previously.

Tragedy in general still conforms to the observations made by Aristotle in the Greece of over two thousand years ago. One such principle is directly relevant to our theme: the central character has to be a recognizable human being, not endowed with god-like fortitude or foresight, but with strengths and weaknesses of character, just like everyone else. His or her downfall (for the Greeks were quite free in their assignment of tragic denouements to female as well as to male characters) is brought about as a result of some apparently inconsequential misdeed or lack of judgement. At the time, the hero pays no attention to it, but the unfolding of the play reveals the devastating consequences of a single act.

Just a single example will suffice. In Sophocles's *Oedipus Rex*, the hero is King of Thebes and loved by all his subjects. Yet, early in the play, in a fit of temper, he killed an innocuous old man standing in his way on the road. By the standards of the time, this was of little import. It turns out, however, that the old man was Oedipus's father, and the play ends amidst the gruesome suicide of Oedipus's mother and the self-imposed blinding and eternal exile of the hero.

This may all seem a long way from the abstract problems of the ability of agents to make decisions in the face of uncertainty, but a key quality that makes literature great is its ability to reflect the human condition. The works of Sophocles and Virgil, Dante and Shakespeare still speak to us over the centuries precisely because they are so emotionally realistic. Compressed summaries of their plots do seem deeply implausible – it is not everyday, for example, that one kills one's father and finds one's mother hanging from the rafters – but the actions and thoughts of the characters, and how they cope with the uncertain consequences of decisions, remain completely realistic.

At a less elevated level, popular contemporary television series such as *ER* or *The Sopranos* could not hold an audience if they were in settings which were too far removed from reality. A key attraction here lies in how the plot will develop, in what will eventually happen to a character as a result of a decision which he or she takes today. Uncertainty of outcome is central not just to great works of literature but to popular entertainment. The characters do not know for certain what the impacts of their decisions and actions will be. If this were not realistic, true to life, the works would not retain their attraction for long.

Conventional economic theory has no such problems about intent and outcome. Its actors, whether firms or individuals, do face constraints on their decisions, but these are of an entirely calculable kind. At its most basic, there are limits to the amount which an agent can spend. This is not regulated completely by the level of income, for there is access to credit in various forms. There is, nevertheless, a constraint on spending. However, within any such restrictions, agents follow a very straightforward behavioural rule: they act so as to maximize their individual utility.

In the standard model within economics, agents have access to complete information. In other words, they know everything that there is to be known about any particular decision. The only thing which can go wrong once the decision has been taken is an external shock which is entirely random. As a result, it cannot be anticipated by anyone concerned in whatever the relevant decision is.

Their omniscience extends beyond the gathering of information. It extends to their ability to process it. No amount of information, no matter how large, daunts the actors in economic theory. They are able to analyze it in such a manner that they are then capable of taking what is the best decision as far as their own interests are concerned: not merely a fairly reasonable decision, not even a distinctly good one, but the very best. We described in the previous chapter how economists try to use game theory to preserve the concept of a single best strategy even in the face of uncertainty.

Armed with these terrifying powers of cognition, both in the gathering and processing of all relevant information, it is hard to see how failure occurs at all. Indeed, the existence of unforeseen shocks completely external to the problem has to be posited. By their very nature, these shocks must be random, for otherwise a fully informed, maximizing agent could, and indeed by definition would, anticipate and hence offset their impact.

In certain restricted circumstances, this way of thinking about the world may give a reasonable approximation to reality. No theory is ever perfect. Perforce, assumptions must be made which simplify reality. The key question is whether the simplifications capture the most important features of the problem being considered. So, with many day-to-day purchases, for example, it might not be unreasonable to postulate that shoppers in a supermarket behave as if they both had access to complete information and could process it efficiently.

By the time the shopper stands in the supermarket choosing between brands of, say, lager or baked beans, a large number of decisions have already been made. In this example, the decision has been made to go shopping rather than, say, to a football match. The decision has been made to go to a food store rather than a clothing store, and so on. The consumer is aware of his or her previous experiences with the various brands, and their reputations. So the amount of information the consumer is required to process in these conditions is small. It is a matter of comparing the prices of the relevant brands. And whether one brand of lager is bought rather than another is not of any great consequence for the future welfare of the consumer.

In these conditions, conventional theory might well be a very good approximation to reality. The demands placed on the cognitive ability of agents are low. However, in many other circumstances, particularly those where decisions have to be made which may have consequences in the future, individuals become far less able to cope with the informational demands required in order to make the very best decision. If Macbeth had simply been deciding what wine to serve with dinner, he may well have been able to arrive at an optimal decision. He was defeated by the sheer dimension of the problem when actions are taken that have important consequences in the future. Neither he nor his wife could anticipate all the possible consequences of their decision.

During the past three decades, a very important advance has been made in economic theory involving a partial relaxation of the very stringent demands placed on the cognitive ability of agents, of their powers to gather and process information. But it is only partial, and the requirements are only altered in a very particular way. Still, it has extended the relevance of theory and expanded the practical domain over which economic theory offers a reasonable approximation to reality. The pioneers of the approach, George Akerlof of the University of California at Berkeley and Joe Stiglitz of Princeton, began working on the issue in the late 1960s and received the Nobel prize for their efforts in 2001. The idea, like many really good ones, appears very simple. Akerlof and Stiglitz relaxed the assumption that agents have access to perfect information. They still process whatever information they may gather efficiently and carry out maximizing behaviour, but they lack complete information. Further, the amount of information that is available to agents may vary. Some may have more than others.

The concept is known as bounded rationality. It applies to situations in which all actors have access to the same amount of incomplete information, and it applies to the more general case in which some have more information than others. For collectors of economic terminology, a rich vein of linguistics in its own right, this latter is known as 'asymmetrical information'.

The relaxation of the assumption that everyone has complete information in any given situation obviously extends the realism of economic theory considerably. John Sutton, currently head of the economics department at the London School of Economics and President of the Royal Economic Society, gives a neat example in his intriguing little book *Marshall's Tendencies: What Can Economists Know?* He spent some time in San Diego around 1980 and, being an economist, took an interest in the city's taxicab industry.*

Sutton's observations have more immediate practical use. While he was in San Diego, the previous policy of restricting the number of cabs by law was abandoned and the number of cabs grew rapidly, more than doubling in the space of a few years. For a time, however, the policy of regulating fares was still in operation, so fares remained high, even though it now seemed that there were too many cabs.

The city authorities then decided to deregulate fares. Surely, they thought, basic supply and demand will work. Prices will fall, given the number of cabs, and those who can no longer make money will leave the market. The price mechanism will balance supply and demand, so equilibrating the market. Unfortunately, this did not work at all, and the reason was bounded rationality. Around 40 per cent of all cab rides in San Diego were taken by visitors. Almost by definition, and unlike the locals, they lacked full information. In such circumstances, an equilibrium involves two groups of sellers. One group of drivers charge a high price and essentially cater to tourists, circling the airport and the main tourist attractions; the other has more business, focusing on everyday use by locals, but charges much lower prices. And this is exactly what happened. Indeed, Joe Stiglitz had published a 'two-price' theoretical model of this kind in 1977.

The consumers in this model are still assumed to process the information available to them efficiently and to take the best, or optimal, decision. The choice in this situation is a very restricted one – presumably one could walk or take public transport instead

* I believe the only other person willing to admit in writing to such a pastime was the eminent Cambridge mathematician G. H. Hardy. He visited his dying colleague, Ramanujan, in hospital in London. Somewhat stuck for words, Hardy mentioned the number of the taxicab in which he had travelled: '1729, a most uninteresting number.' Quick as a flash, Ramanujan replied, 'On the contrary, it is a very interesting number; it is the smallest number expressible as the sum of two cubes in two different ways.'

– and is of no great import as far as the future welfare of the person making the choice is concerned. So the dimension of the choice is low, and it is probably not unreasonable to make the simplifying assumption that agents are able to make the best decision.

However, even in this straightforward context, the concept of a simple equilibrium has to be abandoned. As we have seen, there is not one, but two separate equilibria. Conventional economics is essentially rooted in the Henry Ford philosophy of the early twentieth century: you can have any car as long as it is black. In the same way, orthodox economics, in which agents have enormous powers of cognition, offers a one-size-fits-all approach.

Bounded rationality is the beginning of a more custom-made philosophy. George Akerlof made this explicit in his Nobel lecture, when he stated that 'In this new style [of economics], the economic model is customized to describe the salient features of reality that describe the special problem under consideration. [The standard approach] is only one model among many, although itself an interesting special case.' In other words, the idea of a simple, general approach which will work in every context, or at least most contexts, has to be abandoned. The model must be designed to explain the particular problem at hand.

Much economic theory, however, has barely begun to grapple with the even more interesting and widespread situation in which agents not only lack access to complete information but also lack the cognitive ability to arrive at the 'best' decision. In most real-world situations, it is simply not possible to 'maximize', to find the optimal choice. Reality is far too complicated. Macbeth believed he was acting in his own best interests when he murdered Duncan, but events proved him completely wrong. His decision made at a particular moment had important consequences in the future. And it is precisely when the future matters that the dimension of the problem becomes too vast to tackle by maximization. No one knows the best decision to take.

We have already seen examples of this in the previous chapter with regard to game theory. The Prisoner's Dilemma is a very simple game to describe, but even after decades of intensive research, no one knows the best strategy with which to play the game when the future remains unknown. With no fixed, known end to the game the optimal strategy has not been discovered.

The game of chess offers perhaps a more familiar illustration of this theme. It gives us yet another example of the limits to the ability of agents to process information in an optimal way. Chess is a game in which there is a relatively small number of unequivocal rules, which do not change over any relevant timescale.* The agents – the players in this case – both have full information about the rules, and the position in the game is completely transparent at any point in time, so that a player knows for certain the moves that his or her opponent has already made. Even more, the player knows all the legitimate moves which the opponent could make when it is his or her turn to play. So each player has a huge amount of precise information about the game.

Yet, in most situations in a chess game, we do not know the 'best' move. At the start of the game, it is White to play. Even at this stage, there are no fewer than twenty legal moves available to White. I subscribe to a web-based chess service which holds the details of almost 300,000 games played between international masters and grandmasters. Almost 250,000 of these games were opened with one of two particular moves. Nearly 50,000 began with one of another three moves. So human practice suggests that the other fifteen possible moves are in some way inferior. But we simply do not know, for example, that when computers are able – by some barely imaginable leap in power – to reduce chess to the status of draughts (or checkers in American English), the move 1. Nh3, now regarded as a bizarre opening move, will not be the best that White can make. This opening move occurs only twice on the database, but we do not know whether at some higher, undreamt-of level of play, it will not turn out to be the best.

The complications to which this easily described game gives rise

---

* Hundreds of years ago, the nature of chess was altered fundamentally by some dramatic modifications of the rules. But they have since remained unchanged.

are immense. No fewer than eighteen moves, for example, are recorded on the database in reply by Black to the most popular opening move by White. In turn, seventeen different moves have been tried at master level by White in response to Black's most popular reply to White's most frequently recorded opening. And this only takes us to White's second move in a game which often goes on at this level of play for forty, fifty, sixty or even more moves. The number of possible combinations of moves rises dramatically quickly, in a super-exponential way, with each move of the game.

This property of the game is by no means confined to the artificial world of chess. Let us return for a moment to our example of two superstores competing with each other in a town, even possibly located near each other on the same retail park. One way in which they can compete – and just one way out of many – is on price. Each store will typically carry tens of thousands of different items: different types of product, different brands of the same product and different pack sizes of the same brand. Information on the opponent's prices can readily be gathered. It is rather harder to obtain this than it is to know the moves an opponent has made in chess, but the prices are displayed on the shelves for all to see. The potential combinations of a move that initiates a bout of price competition between the two stores clearly rise extremely rapidly. For example, we may cut prices on, say, thirty particular offers. Our rival could respond on exactly the same set of brands or pack sizes, or could choose a different set of not thirty but, say, fifty or ten or a hundred out of the tens of thousands of possibilities open to him. We then have a choice of response, and so on. Moreover, this game will usually be set in the context of a wider struggle between the two chains of stores at a regional or even national level, so we have to bear in mind, for example, the prices set at other stores within our own chain, lest we cannibalize – a splendid word actually used by retailers – sales from our own stores in nearby locations.

The stupendous number of permutations of moves that a chess game can follow means that we do not know what the result of a game played optimally should be: a win for White, a draw or even, perhaps through some sophisticated Zugzwang,* a win for Black. For aficionados of game theory, it might be thought that chess can be classified as a symmetrical game, in which case theoretical results on the outcome of a game played optimally do exist, even if the actual position which arises from this optimal set of moves remains unknown. However, we do not even know whether the game *is* symmetrical, for we do not know whether the fact that White moves first confers an advantage or not.

In a very limited number of situations compared to the total possible in chess, there will be a single best move, which most players beyond complete novices will find. In rather more situations, but still very few in relation to the total, there will be a best move, but only strong players will find it.

Perhaps computers can help, though it is important to emphasize here that whilst computers have undoubtedly increased our understanding of the game, and can now play at a level at least as high as the best humans, the fraction of situations in which they can find the optimal move is tiny. There are thirty-two pieces in the game, but only when there are five pieces on the board have computers been able to solve all possible situations. Some of these have reversed human belief. Take, for example, the situation that occurs when one side has the King and Queen and the other the King and two Knights: whilst in practice the Queen has very good chances of winning, it is now known that with best play this ending is drawn. The potential number of combinations of moves rises almost astronomically whenever a new piece is added. Some six-piece combinations are now solved. Many of these, one of which involves mate in 243 moves, are believed even by the very strongest players to be completely beyond the capacity of humans to solve. The information is readily available; the constraint to finding the 'best' moves is the ability to process it.

So chess is another example of a game which can be described

* A technical chess term meaning a particular type of position in which whoever is to move is put at a disadvantage.

very simply, but where the dimension of the problem of solving it scales in a super-exponential way. Even very powerful modern computers can only solve a limited proportion of all possible six-piece combinations, yet, to reiterate, the game itself involves thirty-two pieces.

The 2002 Nobel prize-winners for economics, Vernon Smith and Daniel Kahneman, have both worked extensively on situations where agents either do not or cannot maximize. Rather than rely on *a priori* theorizing about how rational individuals ought to behave, they have carried out extensive empirical tests about how people really do behave. In so doing, they have created an entirely new field known as experimental economics. In the same way that computers have reversed previous beliefs about the outcome of chess games with certain combinations of pieces, experimental economics has altered, sometimes quite dramatically, our view of how humans actually behave.

Kahneman in particular is scathing about the maximization hypothesis on individual behaviour. In his Nobel lecture, published in the *American Economic Review* in December 2003, he dismisses rational models as being 'psychologically unrealistic'. He has especial scorn for the standard approach in economics, describing the theory in which people make choices in a comprehensively inclusive context as 'particularly unrealistic'.

Economists have been willing to accept agents who lack perfect information. They have even begun to accept, just about, that in many situations agents may not be able to carry out maximizing behaviour. But the last taboo seems to be the idea that, even if agents do not or cannot gather complete information, even if they do not or cannot carry out maximizing behaviour at a given point in time, they learn. They learn a better strategy, which may even begin to approach a maximizing strategy.

We can think of this in the practical context of the game of chess. How much learning is feasible here? The environment is fixed, for the rules of the game do not alter over time. Information on the opponent is transparent, for his or her moves are known with certainty once they are made. Many of the features of the real world that make the prediction of outcomes even more uncertain for decision-makers in the public or private sectors are not present in the game of chess. In reality, the environment is not fixed, with existing players being free to invent new moves and new players altogether being able to enter the game and knowledge about the opponent may be difficult or almost impossible to obtain.

Even in the context of the game of chess, when many of the real-life barriers to learning do not apply, learning is very limited. We can distinguish two types of learning in the game of chess: learning by an individual agent, and the collective learning due to the efforts of all agents, shared via books or articles. The former can, of course, refer to the latter, but it helps to discuss them separately. An individual agent will (usually) learn quite rapidly to avoid catastrophic blunders. Beyond this, learning will differ widely amongst agents, either because of the degree of interest or because of limits to the natural ability of the agent. We also have collective learning, which undoubtedly takes place. The strongest chess players today, in so far as we can judge this, play at a level which is considerably higher than players, of, say, 1800, and probably higher than those of 1970. It is hard to imagine that this is due to some sort of genetic mutation. The leading players of recent decades, such as Bobby Fischer or Garry Kasparov, for example, are fulsome in their praise of the inherent abilities of some of the great players of the nineteenth century, believing them to have had the potential to play just as strongly as the modern-day giants.

The increase in strength which has taken place is attributable to advances in collective knowledge. Games played by the strongest players are in the public domain, to be analyzed by others at leisure, who publish the results of their thinking in newspaper and magazine columns. Books on various aspects of the game are written by leading players to codify our knowledge of it. However, the timescale over which learning takes place seems to be long. Can we meaningfully say that the strongest player in 2004 is stronger than the one in 2003 or even 1994? Not really. Progress is measured in decades rather than single years.

Also, just how much learning has actually taken place in chess?

As we have already noted, most of the positions which occur in chess still do not admit of a single, unequivocally best move, even when analyzed for hours by modern computers. Most of the time, it seems that even very strong players make 'rule of thumb' moves. The writings of players such as Kasparov make this clear. They do attempt to foresee the future, to calculate the consequences of any given move which they make now. In doing so, they will attempt to anticipate the possible responses from their opponent. But, in most situations, it is literally impossible for them to calculate all the possible variations.

Instead of a futile search for the best possible move, most of the time chess grandmasters use their skill and experience to make what they consider to be a reasonable one. They make moves that seem good and that avoid obvious loss, exactly in the way experimental economics leads us to believe individuals and firms behave.

Somewhat paradoxically, economists have long been familiar with a problem in which it is not possible to optimize, to find a single best strategy. The context is at the very core of free-market theory. In financial markets, the efficient markets hypothesis states that it is not possible to construct a strategy based upon the past behaviour of asset prices that will consistently outperform the market as a whole. No matter how much effort is expended, no matter what techniques are used, this result will hold. In other words, it is impossible for agents to learn.

In such a situation, a purely random choice of portfolio is likely to do as well, or as badly, as one chosen by an 'expert'. Of course, this does not mean that the price of equities is literally determined at random. On the contrary, there will be a motive and a rationale behind every single trade. Instead, we should think of the determination of asset prices as being a very high-dimensional problem. So many factors can influence the price of an asset – and the relative strength of these factors almost certainly varies over time – that the analysis of the data becomes intractable. We have all the available information, but there are limits to our ability as agents to process it.

This is the economics of the twenty-first century. Individuals, firms, governments, households may lack access to complete information. Even more importantly, they do not have the cognitive ability to process it in a way which finds the single, optimal choice. Particularly when confronted with decisions that have consequences in the future, the problem of finding the 'best' move, the best strategy, is simply too hard. Instead, agents look for reasonably good strategies which avoid obvious loss, and they find it very difficult to learn better strategies. Armed with this view of the world, we return to the problem of failure and extinction.

7 'The Best-Laid Schemes ...'

The widely used quote which is the title of the chapter is taken from the poem 'To a Mouse', by Robert Burns, the most famous writer of Scots English and Scotland's national poet: 'The best-laid schemes o' mice an' men/Gang aft a-gley.' The *New Dictionary of Cultural Literacy* puts this rich, original phrasing in two different ways: directly, as 'the best-laid plans of mice and men often go awry'; and in a more prosaic way: 'No matter how carefully a project is planned, something may still go wrong with it.' By now, we should be becoming accustomed to this sentiment.

Running a business or a government is much harder than playing chess. Of course, some play these roles better than others, just as some are better at chess than others. But there are fundamental differences in the order of difficulty involved in the two tasks, no matter the level at which they are tackled, and it will be helpful to examine these in some detail.

As we have seen in the previous chapter, the ability of humans to comprehend chess is very limited. Only a tiny fraction of all possible positions can be understood completely. Most of the time, the moves of even the strongest players may have unforeseen and unpredictable consequences. Given that this is the case with chess, it is even more true when it comes to the decisions of company directors or government ministers.

One obvious difference is that in chess the rules of the game are fixed. The eccentric Bobby Fischer, arguably the greatest player in the history of chess and undefeated as world champion until his self-imposed exile from the game, became a great enthusiast for random chess. However, even these changes to the game were strictly limited: the pieces kept the same powers from game to game. It was just their starting positions that were decided at random. Imagine how much more complicated and difficult the game would become if the actual rules of how the pieces move were also changed at random, not just from game to game, but during the course of any particular game.

The real world of decision-making corresponds much more closely to the latter than to the fixed-rule environment of orthodox, or even Bobby Fischer, chess. There are limits on what both the player and his or her competitors can do while still remaining within the law. Otherwise, there are no holds barred. Competitors can try different strategies, just as in chess, but the variety of these is limited not by prescribed rules, but simply by their imaginations. In particular, new products can be introduced which threaten an incumbent in previously unexplored ways. IBM, for example, dominated the world of computing until the personal computer came along. This changed the rules of the game dramatically and almost brought IBM to its knees. Not just a different way of moving, but a completely new piece with its own new ways of moving was introduced to the game.

Chess is also simplified dramatically by the fact that there is just a single, known opponent in any particular game. More importantly, the opponent does not change during the course of a game. At any point in time, a business will usually be aware of its main competitors. Typically, however, there will be more than one, each with its own moves, its own strategies. The company has to respond to not just one opponent but several.

Even so, it is not always obvious who the competition is, and companies spend a great deal of money trying to understand this better. Consider, for example, a whisky producer with its own range of brands. The direct competition for, say, its medium-priced brand will undoubtedly include similar brands made by other companies. But what else is in its competitive set? There is a considerable list of possibilities: both more expensive and cheaper whiskies are obvious candidates for consideration. Many consumers, however, may be looking not for a whisky as such but for a product in the same price range which also delivers a substantial amount of alcohol, such as gin or brandy. Moreover, even when a company has decided what the competitors are at any point in time, it must try to take account of the fact that consumer tastes are not necessarily fixed over time. Further, a new competitor may suddenly enter, such as Alcopops, a bottled mixture of soft drinks and hard spirits, which experienced tremendous if transient popularity in Britain in the late 1990s.

A further crucial difference between chess and real-life decision-making is that chess has a single, unequivocal goal: you win by checkmating the opponent's King. This is the sole purpose of strategy. There is no other way of winning. In business, defining the criteria for success is not completely straightforward. Businesses exist not to perform acts of charity but to make profits. Of course, at an even more basic level, they need to survive. This much is easy to define but, once we introduce time into consideration, things become more difficult. A tactic or strategy that might make large profits quickly is not necessarily the best in the longer run. Pricing a product or service very high might yield dividends in the near future but might also cause customers who would otherwise have been loyal to look elsewhere.

It is in government rather than the world of commerce where goals are particularly multifarious and not easy to define. Very occasionally, a government has the luxury of being judged on a single outcome. At time of war, for example, your country may emerge the victor or it may be defeated. In the Second World War, the sole purpose of the British and American governments was to defeat the Germans and Japanese. This gave them carte blanche to introduce measures which would otherwise have been unacceptable. In Britain, civil liberties were to all intents and purposes suspended and many innocent people interned. In the US, private property was effectively commandeered by the state. Following Pearl Harbor in December 1941, the production of domestic automobiles was forbidden by government fiat. In a stupendous feat of organization, the assembly lines in Detroit were reconstructed so that by as early as February 1942 they were given over entirely to military production.

In the 1930s and 1940s, the Soviet Union was able to carry out a massive programme of industrialization. With simple goals defined by successive Five-Year Plans, and supported by a ruthless political police, the government was able to succeed. The hardships and degree of inequality inflicted on the population far exceeded that of the industrialization period in the west a century or so previously, to say nothing of the mass famines and labour camps. Indeed, in the early 1950s the diet of the average Soviet citizen was scarcely any better than that on offer in the camps, but a huge primary manufacturing economy based on coal and steel was built. Even within a planned economy and one-party state, where dissent could be and was suppressed, the transition to a successful consumer-based economy was never carried out. Supplying a wider range of more complex products to consumers, to meet more complicated goals, proved far beyond the ability of the system.

In general, it is rare for a government to have a single criterion of success. So many competing and conflicting goals exist. The real-life equivalent of checkmate remains a chimera. Politicians are sometimes beguiled into believing that, in Bill Clinton's phrase, 'It's the economy, stupid', but this by no means always works. The Democrats were defeated in 2000 despite having presided over an almost unprecedented upsurge in American wealth and living standards. In Britain, the Conservatives were re-elected in 1983 and in 1992, even though these years followed immediately after what had been by far the worst economic recessions in the UK since the 1930s. Millions of people had lost their jobs, yet the government was re-elected.

In the real world, the environment that a firm or government faces is constantly changing. In the commercial world, competitors are always thinking up new strategies. New competitors can and do emerge out of the blue. There are examples of this all the time. For example, Unilever, a giant company that might be expected to be able to understand its market, paid over $1 billion for Slim-Fast, a company that makes disgusting but successful diet drinks. It must have seemed like a gold mine. People in the west are getting fatter and fatter, and at the same time more conscious of the need to lose weight. Sales of Slim-Fast products were booming but, very soon afterwards, the Atkins diet, based upon massive consumption of protein, swept many countries, and Slim-Fast sales collapsed.

Microsoft is a corporate giant. Surely if any company epitomizes the rational, calculating strategy of success, it must be Microsoft. Yet the reality is wholly and completely different. An excellent account is given in Marlin Eller's book *Barbarians Led by Bill Gates*. Eller was from 1982 to 1995 Microsoft's lead developer for graphics on Windows. Windows now, of course, dominates the PC operating systems world, but its success was far more the result of a series of accidents than of a far-sighted, planned strategy.

Eller's introductory remarks are worth quoting at some length: 'There was a great disconnect between the view from the inside that my compatriots and I were experiencing down in the trenches, and the outside view ... in their quest for causality [outsiders] tend to attribute any success to a Machiavellian brilliance rather than to merely good fortune. They lend the impression that the captains of industry chart strategic courses, steering their tanker carefully and gracefully through the straits. The view from the inside more closely resembles white-water rafting. "Oh my God! Huge rock dead ahead! Everyone to the left! NO, NO, the *other* left!" ' Eller goes on, 'Reality is rarely a simple story and is probably more like a *Dilbert* cartoon.'

We are back in the world of Bob Lutz, overall product chief at General Motors, whom we met in Chapter 2. Commenting on Japanese plans to enter the lucrative American pick-up truck market, he said: 'The risk for us is if consumers prefer Nissan styling and their power trains.' In other words, GM did not really know whether their strategy would be successful in fending off a new Japanese invasion of yet another sector of the American automobile market.

Another corporate giant unable to penetrate the veil of uncertainty shrouding the future was Ross Johnson, CEO of the giant RJR Nabisco corporation. In the late 1980s, Johnson was at the height of corporate power. Based on the tobacco products of R. J. Reynolds and the consumer snacks of Nabisco, the company appeared almost to run itself. Johnson had all the time in the world to be flown in company jets for long weekends in exclusive Caribbean retreats, to the best golf courses to play with showbiz luminaries, to indulge almost every whim on expenses, in addition to the salary, bonuses and pension contributions which poured into his bank accounts and portfolios.

He was persuaded that he could make even more money, stupendous amounts, by executing a management buyout. The company's shares had been trading at around $45 for some time. By making a bid of over $70, he was certain both that the bid would be accepted and that he could go on to make the company even more profitable and hence, as the owner, generate vast wealth for himself. We might pause to ask why it was that he, in common with many CEOs, did not exert himself already to increase profits for the shareholders, or indeed why the rest of the board did not realize that this was feasible. However, his management buyout bid attracted the attention of predatory Wall Street firms, and in particular Henry Kravis's KKR company. After a thrilling saga of bid, counter-bid, tears and tantrums, KKR was eventually successful at a share price of well over $100. This leveraged buyout was valued at $25 billion, making it the largest even in the frenzied period of late 1980s Wall Street.

So a CEO of a giant company at the height of his powers misjudged entirely the consequences of his attempted management buyout. He thought he would succeed. He failed. He thought the company could be bought for some $70 a share. It was eventually sold at over $100.

But to return to the chapter and verse of the Windows story. In the late 1980s, the main strategic goal of Microsoft was to link up very closely with IBM. In particular, the two companies were developing jointly a new operating system, OS/2. Windows merely limped along. Bill Gates staged a major publicity coup at the computer industry's biggest exhibition, COMDEX, in 1983. He announced that Windows 1.0 would be shipped in the spring of 1984. After immense effort, it finally appeared in November 1985. The reviews were blisteringly awful. The product's size was huge relative to the capability of the personal computers which existed then. The *New York Times* observed that 'Running Windows in 512K of memory is akin to pouring molasses in the Arctic.' In Eller's blunt description: 'The product was essentially useless.' The support team within Microsoft for Windows was cut back to a mere three people.

In contrast, great effort was being put into the relationship with IBM. In October 1988, the two companies launched OS/2 Presentation Manager, with Bill Gates proclaiming, '[This] will be the environment for office computing in the 1990s.' Marlin Eller quotes Steve Ballmer, Gates's number two, as saying, 'This is it, after this we're not going to have any more Windows. It's all OS/2.'

Windows 2 meanwhile had been launched, with little success. Only a couple of people were left within Microsoft to maintain the product. Sporadic development of the product still took place on the next version, Windows 3.0, but an article in the *National Review* summed up the view of the industry: 'Microsoft would cease development of its Windows software after the release of Windows 3.0 ... IBM's OS/2 would become the main PC operating system for the 1990s.'

On 22 May 1990, Windows 3.0 was made available to the public. It sold two million copies in the first six months.

The point of this story is not in any way to denigrate Bill Gates and his team. Far from it. They showed great flexibility and the ability to seize very rapidly upon opportunities which were presented to them. They took the correct view that the personal computer market would grow massively. Their vision was to have Microsoft involved in software which would become the leading choice of consumers in this market. The point is that, despite the enormous business abilities of Gates and his key players, they did not foresee that it would be Windows and not OS/2 which would fulfil this role. Windows was almost abandoned as a standalone product. Its support team was cut to virtually zero. And yet it proved a massive, overwhelming success. Success, like failure, comes in many guises.

It is the sheer complexity associated with many decisions that defies the orderly application of the rational calculations of economic theory. The number of possible permutations of outcomes is simply too great to be computed. The degree of uncertainty rarely permits the computation of the optimal, the unequivocally best strategy at any point in time. Even in chess, this is possible in only a tiny fraction of all possible positions, and, as we have seen, running a firm or a government is much harder than playing chess.

We have also already seen how, even in apparently very simple theoretical models, the ability to find the 'best' move rapidly becomes beyond the capability of the players. In Chapter 4, the Schelling model of residential location exhibited this property. Each player has a preference for the neighbourhood in which he or she would like to live, but the outcome for any individual is so contingent upon the future actions of others that it is not possible to compute his or her best choice at any point in time. Similarly, the Prisoner's Dilemma game in general permits no clear-cut solution in terms of the choice of the 'best' strategy.

There are many examples of theoretical models such as the Schelling model or the Prisoner's Dilemma which have similar properties. As interest in the new form of economics grows, the number of models of this kind is also growing rapidly. In the new economics, we not only address a specific problem, we try to start from the outset with rules of behaviour which have empirical support rather than with rules which we believe *a priori* a rational agent ought to follow. John Nash criticized the RAND analysts who played Prisoner's Dilemma for not playing his rational, Nash equilibrium strategy. His theory dictated that this was how people ought to behave. If they didn't, it was the people who were at fault and not the theory. This attitude still pervades much modern economics. However, when we look back over economic theory, there are nuggets of gold amongst the dross. One of these, a model with a clear practical application to political and business strategy, has been known to economic theory for over seventy years. It, too, is simple to describe but fiendishly difficult to play in an optimal way, except in very restricted circumstances.

Harold Hotelling was instrumental in creating the statistics department at Columbia University in the early 1930s. He was responsible for some really major advances in statistical theory, which are still the foundation of a great deal of practical work today. Hotelling made only a very few ventures into economics. One of these was in 1929, during his brief spell at Stanford before joining Columbia. His paper, published in what was then probably the world's leading academic economic journal, the *Economic Journal*, has the rather forbidding title 'Spatial Competition in Duopoly', but it was illustrated with a much more homely example.

Imagine a crowded beach at the height of summer, in the days before the great success of capitalism in the second half of the twentieth century. The diversions and amenities available to the beachgoers are few. Two rival ice-cream sellers are deciding whereabouts to locate on the beach. They know three basic facts: first, the bathers are spread completely evenly across the entire range of the beach. This much is obvious from simple inspection. Second, each person on the beach will at some point during the day want an ice cream, although no one will buy more than one. On a hot day, with people eking out their money over the week of their holiday, this seems plausible. Third, no matter where the sellers locate on the beach, everyone will still want an ice cream. This is less obviously true, but suppose for the moment that it is.

Where should the rivals choose to set up their stalls? Before answering the question, we can see quite readily that the model is capable of generalization to a wide range of different contexts. We can think, for example, of a market containing two very similar but rival brands, each distinguished from the other by virtue of a single characteristic. If we again assume that consumers are evenly distributed in their preferences for this characteristic, whatever it might be, whereabouts along the line of preferences should the offer be made? Think, for example, of two local radio stations, who each offer an identical mixture of music and chat shows. They differ only in the political slant of their news. More generally, in terms of politics, we can readily imagine two competing parties deciding where to make their offer along the ideological spectrum from left to right.*

We are already, of course, making quite dramatic simplifications of reality. For example, the producers, the ice-cream sellers, are assumed to know a great deal about consumer tastes and preferences. They know that everyone will want an ice cream, no matter where they locate. So they know the size of the total market, and the task of each of them is to get the largest share of it for him or herself. If we think in political terms, say, the size of the market is the percentage of the electorate who turn out to vote. The number of abstentions may be crucial in deciding the outcome. In the British general election of 1997, by way of example, the Conservatives suffered an absolutely devastating defeat, after having been in power since 1979. All the polls suggested in advance that they would indeed lose, but the scale of their defeat took the pollsters by surprise. Many Conservative supporters proved in the event to be so disillusioned that they did not bother to vote on the day, and the turnout fell sharply. People failed to predict accurately the total size of the market.

We need one more assumption before we can put the model to work. We suppose that the products of the two ice-cream sellers are absolutely identical, and the only thing which distinguishes them in the eyes of the consumers is where they are situated on the beach. Everyone will buy from the seller located closest to them. In practice, of course, even a simple product like an ice cream may be differentiated in numerous ways: price, flavour, quality and the like. But, to keep things as simple as possible, we assume away these complications.

We can now imagine ourselves as being in the position of extremely highly paid consultants commissioned to give advice to two giant multinational companies on where to locate in a new

* Of course, in practice the extremes may almost join on a circle rather than be far apart at the ends of a line. Authoritarian fascism on the far right and Trotskyism on the far left have much more in common with each other than they have with those who inhabit the middle ground.

market. We may even fantasize that our fees will be uplifted*
depending on how much of the market our firm actually obtains.

Of course, we know that Hotelling's model makes many simplifying assumptions that may not hold in practice, but these are
complications that can be taken into account later. If, by some
means, we know that only two firms are intending to enter the
market, the model in its very simplest form enables us to offer very
precise advice to the ice-cream sellers about where they should
each locate.

Figure 7.1 below shows the consumers spread out along the
beach at regular intervals. For descriptive purposes, assume the
beach is 100 units long.

Markets are meant to serve customers, and from the customers'
point of view it would be best if the sellers set up stalls at the
points marked 25 and 75. In other words, a quarter of the way
along from the furthest left of the line, and a quarter of the way
back from the furthest right. At these locations, the travelling time
for the group of consumers as a whole is at a minimum. No one
has to travel more than 25 units to get an ice cream. If the firms
located at, say, 10 and 90, then the customers at point 50 would
have to travel 40 units. A moment's reflection will show that at any
combination of locations other than 25 and 75, the customers will
have to travel further.

This outcome appears to be satisfactory for the ice-cream sales-
men as well. Each vendor will get half the total sales, as half the
customers will find themselves nearest to just one of the sellers.

However, a moment's reflection will convince us that this solution is not a stable one. Suppose the sellers do start off at the

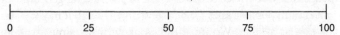

| 0 | 25 | 50 | 75 | 100 |

FIGURE 7.1 Customers evenly spread along the beach in the Hotelling
location model/game. Locations designated by points between 0 and 100

* A marvellous word that I have only ever heard lawyers in the City of London use
with this particular meaning. It is self-explanatory, and at the same time preserves
their gentlemanly mystique by avoiding vulgar commercial terms such as 'success fee'.

25 and 75 combination of points. They observe the patter of feet and procession of customers to their stalls. It might quickly occur to the vendor at point 75, say, that by trundling his cart just to the right of point 25, he could capture almost three quarters of the total market. All the consumers to the right of his or her new location will buy from there. But then the seller at point 25 could jump just to the right of where its rival has moved, and so on and so forth. In more realistic examples there may, of course, be substantial costs associated with a move. In particular, it may be difficult to reposition a brand, whether in consumer markets or politics, too much. It is a very tricky task to retain the loyalty of a sufficient number of existing consumers and at the same time make the offer so different that it taps into a new part of the market entirely. We will see, however, that even keeping out such complications rapidly leads to very complex situations in the Hotelling location model.

From the perspective of the sellers, as they chase each other along the beach, the unequivocally best location for both of them is exactly halfway, at point 50. Each will continue to get half the market, as they would at points 25 and 75. But in this case neither can outmanoeuvre the other to a superior location. The customers may not be best pleased, because their collective travelling time has increased. Those at the very end of the beach, for example have to travel 50 units of distance, whilst half of the total number of customers now have to walk more than 25 units, the maximum distance anyone has to go at the original 25 and 75 locations. But if they want an ice cream, there is nowhere else to go.

So in this case we can happily collect our fee, safe in the knowledge that our advice to locate right in the middle generates the maximum possible sales for our company. Though it might not be as easy as all that. We are already making many simplifications, some of them quite drastic, which may well not hold in practice. Any real-life decision is therefore bound to be more complicated. Suppose, for example, we allow consumers to start to take into account the costs of travelling to a location. The further a customer has to travel, the greater the inconvenience. Again, this applies not just to physical location. In a political contest, for example, in which there were just two mainstream parties each with a centrist offer, voters with more extreme views may not bother to vote at all. The effective distance between their own preferences and what is on offer means that they do not participate in this particular market.

We can invent a scenario in which, unlike when the distance travelled by the customers isn't a factor, it is best for the two competitors to locate at the points 25 and 75. Suppose we know for certain that customers who are more than a certain distance away, whether in physical location or in preference for what is on offer, will not buy at all. For simplicity, say this is 25 units of distance. Still keeping the arithmetic simple, the beach is 100 units long. Following the analysis of the initial model, the two firms initially locate right in the middle, both at point 50. One gets all the custom from people between 25 and 50 units along from the extreme left-hand end of the beach, and the other gets the next 25. The idle people at either end of the beach cannot be troubled to waddle along for an ice cream. The total market size is fifty customers, and each seller gets twenty-five each.

Now, however, one of the sellers could move to point 25. In this way, he gets everyone situated between 0 and 25. If the rival stays put at point 50, they divide between them the customers between points 25 and 50, and the customers between 50 and 75 all go to the competitor at point 50. They still split the market equally, but its total size has expanded. By the same logic, the competitor discovers that an even better location is at point 75. At these two locations, they sell to everyone. No one is further away than 25 units of distance from an ice-cream stall and, as it happens, these are the best locations for the customers.

Equally, however, we can construct scenarios involving yet another set of assumptions on consumer willingness to travel which will lead to outcomes in which the rivals locate towards the extreme ends of the beach. The logic is by no means as straightforward as it is in the two examples above. In fact, it is rather complex. Claude d'Aspremont, Jean Gabszewicz and Jacques-Francois

Thisse discovered this result and published an article on it in the leading journal *Econometrica* in 1979. Essentially, they examined what would happen if consumers perceived that there were costs to them in purchasing a product somewhere other than at their own precise location, and that the costs rose more rapidly than the distance. In the context of the ice-cream sellers, the costs are simply the effort and time it takes to walk to them. More generally, we can think of the concept of distance for any individual wondering what brand to buy as being the difference between the various products which are actually on offer and his or her ideal brand.

The exact outcomes depend upon the particular assumptions that are made about how consumers see the costs associated with purchasing something which is either not at the exact location where they are or which is slightly different from what they want. But in general, in this kind of scenario, the two firms will usually locate towards the two extremes.

So, we continue to assume that there is only one quality which distinguishes the two rival products; that there are only ever two firms in competition with each other; that consumers are spread completely evenly in their locations across the possible range of preferences; that there are no costs involved in the firms relocating. In short, we are still making drastic simplifications about reality.

But already we can generate outcomes in which the best locations for the firms are either in the middle at point 50, at the equidistant points 25 and 75 or towards the extremes of 0 and 100. To differentiate between these outcomes, we need some very precise knowledge on the costs that consumers think they incur if the offer is not made at their exact location or preference. Perhaps our consultancy company is part of a multinational group, and we are able to recommend a market-research firm to our client in an attempt to discover this information. Even so, as we have seen with the examples of Nissan, GM, Chrysler, Microsoft and others, it is by no means an easy task to discover consumer preferences. Mistakes are often made.

Things become even more difficult to disentangle when there are more than two firms involved. A particularly important piece of information which can affect the outcome dramatically is whether the firms are assumed to enter the market simultaneously or sequentially, a complication which we have previously ignored. Do all enter at the same time? In certain circumstances, such as a government auction for cell-phone bandwidth, the answer is yes. But, in general, this will not be the case. Even in a completely new market, some will try their luck at an early stage, whilst other potential entrants will wait and see, hoping to take advantage of the early movers' mistakes. Again, in practice, firms will not know for certain how many others actually will try to enter the market. They will have a reasonable idea but, short of illegal industrial espionage, they can never know for sure.

It might seem that the best strategy, or at least a very good one if there are substantial costs involved with repositioning or relocating the offer, is to wait until others enter the market and then choose the best spot with this information to hand. The early entrants may, however, attract customer loyalty. It might prove difficult to entice the customers away with a different offer, even if it is one which in terms of its objective qualities is better. Is the QWERTY keyboard, for example, the most efficient one ever invented? The answer here is an unequivocal 'no'. Yet it retains a lock on the market because it is the one to which almost everybody has become accustomed. Is Windows the best operating system ever discovered? Perhaps I had better just say that everyone has his or her own opinion, and leave it at that.

Suppose we restore the assumption that all consumers in the range from point 0 to point 100 will make a purchase, regardless of where firms locate. Consumers are still spread evenly along the range from 0 to 100, and differentiate between the rival products only on location. They will always buy from the nearest offer, exactly as in the very simplest version of the Hotelling model discussed above. Let's now make the assumption that it is too costly for firms to change their initial location. This is not completely true in practice, but it is a much more realistic assumption than the one which asserts there are no costs at all involved. As we have noted, it is very hard for a brand to change its image.

The two-firm example remains easy to understand with this set of assumptions. We advise our client to locate at point 50, and doubtless the competitor's advisors do the same. Again, we collect our fee.

Now, instead of just two entrants to the market, imagine that there are five firms involved. This appears to be a manageable number and is realistic in many actual markets. This simple increase brings enormous complications.

Everyone involved can be presumed to know the best location when there are only two firms involved, because they are all paying high-quality advisors such as ourselves. So why not try the same strategy and locate slap bang in the middle at point 50? An immediate problem is that, if we think everyone else will do this, we will do much better by locating just to the left or right of this point, at 49 or 51, or 49.9 or 50.1, depending on how finely we can grade the market in practice. When we have just one rival, this does not make sense. When we are both at 50, we both get half the customers. If I move to point 49, I still get everyone between 0 and 49, but I lose half of those between 49 and 50. Perhaps not many, but still less than at my best location. If five firms are involved and we are all at point 50, we each get one fifth of the total number of customers. By locating at 49 or 51, I now do better than at 50, rather than worse.

If we follow this train of thought and imagine that rival advisors can also work this out, perhaps we should suggest locating at point 48, which we hope will outflank them. But if a competitor actually goes to 47, say, and another to 49, our client will do very badly indeed.

We can in principle work out our best location for every possible combination of locations of our four rivals. A slight problem is that, even if firms are only allowed for some reason to locate at integer values of points (the whole numbers, 0, 1, 2, 3 and so on), there are almost four million possible combinations. Customers in practice may not be able to differentiate between positions which are very close together but, if we grade the market more finely and permit a distinction to be made between locations involving just the first decimal point (such as 0.1, 0.2, etc.), the number of potential combinations is over forty billion. A somewhat more serious problem is that only one of these combinations of locations of our rival firms can actually happen, and at this stage we do not know which one it might be.

This may all seem very complicated. And it is, but now it is about to get worse.

A useful way to proceed in circumstances such as these, where the application of mere human logic might not take us very far, is to solve the model on a computer and work backwards, as it were. In other words, we set up the model on a computer and assume that firms choose their locations at random. We obtain a large number of individual solutions of the model on the computer. It is as if we can rerun history many, many times, with each individual set of locations for the five firms and the resulting market shares associated with them being one possible historical event. We are not going to run so many solutions that we cover the full range of the four million, forty billion or even more possible outcomes but, provided that we run a few thousand or even a few hundred, say, we can rely on probability theory to tell us that we have a sample which is representative enough to inform us about the various combinations of locations that might occur. We can then analyze the outcomes of these hypothetical histories and see if we can learn anything from them. For example, do firms that locate near the centre at point 50 tend on average to do better than those which place themselves near the extremes? If this turns out to be the case, we can give advice on this basis.

We would have to be much more circumspect than if there were only two competing firms. With the particular set of assumptions we are using about customers' willingness to travel and so on, we could with two firms be absolutely confident that the best location is at point 50. With five, we might at best be able to say that on average a firm will do better at point 50. That is, if we could rerun history many times, this location might give a market share which on average is better than any other. There is no guarantee at all that in the one actual history we experience this will turn out to be the case. It may prove a disastrous choice, but at least we should give ourselves the best chance prior to the event. If, for some bizarre reason, we were forced to stake our lives on the toss of a coin, and we knew that the coin was slightly biased towards heads, it would be a very brave or foolish person who would call tails. Your life may still be forfeited if tails does indeed come up, but calling heads gives you the better chance to live.

The results of 200 separate solutions to this version of the model, in which five firms locate at random, are plotted in Figure 7.2. On the left-hand axis, we plot the market share achieved by a firm, and on the bottom axis its position. Each point in the chart corresponds to the market share obtained and the location of a particular firm.

Firms end up with very different shares of the market, the lowest being just 1 per cent and the highest 69 per cent. There is no absolutely obvious pattern to the relationship between share and location in Figure 7.2, but closer inspection suggests that locations at the extremes, near 0 and 100, do not seem to do as well on average as other locations. They do not often have a high market share. We can easily confirm this by calculating the average share

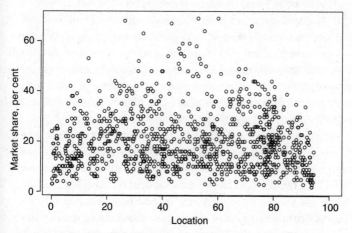

FIGURE 7.2 Five firms entering a market in the Hotelling game, with only one feature differentiating the rival offers. Consumers are evenly spread between point 0 and point 100. (Two hundred solutions of the model.)

obtained by firms locating between 0 and 10, between 10 and 20, and so on. Even so, there are examples of firms locating near to 0 and 100 who happened to get a 40 per cent market share.

It is much less obvious that there is an advantage to being closer to 50 than, say, 30 or 70, but if we now actually earn our fees and do a little bit of statistical analysis, we can say that there is a very slight tendency for firms in the middle (around point 50) to do slightly better than at other locations. So it seems that we should go along with the best guess of the marginally biased coin and try to save our life, or in this case our fees, by locating at 50 after all.

There is, however, some far more decisive evidence we can draw from this game: the critical determinant of success is not the absolute position of the firm on the line of consumer preferences; it is the distance of its offer from its two nearest rivals.* In other words, the more distinctive your offer, the better you do. This conclusion has the advantage, to some at least, of seeming to go along with common sense.

We can see this in Figure 7.3, which plots market shares obtained in the two hundred solutions to the model, or game, and the firm's distance from its two closest competitors.

We do not need any statistical expertise at all to understand this chart. Our ideal advice should be: 'Locate as far away from the nearest competitors as you can.' But a moment's reflection will tell us that, short of insider knowledge, this is a virtually impossible piece of advice to follow. The number of possible permutations is so large as to become very hard to handle. Suppose we place the offer at point 0.01, in other words, very close to point 0. We will do well if everyone else locates above point 50, for the whole of the market from point 0 to point 25 will certainly buy from us, and we might get even more, depending upon how far above point 50 our rivals are. In such a case, our two closest rivals are a long way away from us. But we only need one firm to locate close to 0 but just to the right of us to do very badly indeed. If someone enters at

* The lowest and highest positions are bounded by 0 and 100, respectively, so their distance on this measure is their distance between whichever of these points is relevant and their closest rival.

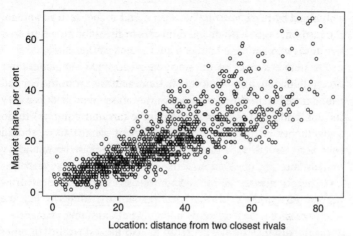

FIGURE 7.3 Five firms entering a market in the Hotelling game, with only one feature differentiating the rival offers. Consumers evenly spread between points 0 and 100.

0.02, say, we will get everyone from 0 to 0.01, and half of those between 0.01 and 0.02. And that is all we will get.

Things become even more complicated if we assume that all potential entrants are armed with the information supplied by Figure 7.2, that on average you may do slightly better by entering near the middle. If we think everyone else will act on this information, then we will do best to locate at one of the extremes. But what if everyone else thinks that everyone else will enter near the middle, and so everyone actually decides to locate near 0 and 100? Clearly, we can get involved in never-ending layers of complications. I think that my rivals think that I think that my rivals think, and so on and so on, with no end to the sequence. And the discipline of psychology shows quite clearly that there are severe limits to the ability of humans to cope with such increasing levels of complexity. Beyond five or so such layers, we simply lose the plot.

As a good approximation, the best piece of advice we can give is: fill an urn with balls numbered from 1 to 100. Shake up the urn, close your eyes, choose one, and locate at that number. In short, choose purely at random. We might, if we know that there is only a small number of potential competitors, be able to do ever so slightly better than a purely random choice by locating at point 50, but this is very far from being an automatic guarantee of success.

The indeterminacy of the solution even in this simple model is certainly known to economists. Again quoting from the Royal Economic Society President John Sutton, 'In analysing the entry stage of the game, we find that multiple equilibria are endemic … we arrive at the point where the search for the true model becomes futile. The problem is that there are many "reasonable" models and to choose one over another we would have to have access to knowledge about various subtle features of the entry process, such as the belief that each firm had regarding the way in which its choice of location would or would not affect its rivals' decision making.'

Running a business is, as we have seen, much harder than playing chess. It is also much harder than analyzing the Hotelling location model, or playing the Hotelling location game, as we might think of it. Even making dramatic simplifications of reality, we cannot really draw any definite conclusions from this game unless we are in possession of immense detail concerning consumer preferences and, even more difficult, the intentions of our potential rivals. And if we cannot make firm recommendations or give clear-cut advice to players of this game, how can we possibly hope to do any better when advising firms who face this problem in real life?

Earlier in the chapter, we saw the consequences of operating in a massively uncertain environment. Unilever, RJR Nabisco, GM and Microsoft all face the same virtually impenetrable veil over the future as the smallest one-person firm. The giants are less likely to make elementary errors of operation than small firms, but otherwise there is no difference. The possible permutations of outcome are vast. Firms and governments certainly act with intent; they intend to succeed. But success often arrives purely by chance. And so does failure.

A great deal of human strategic decision-making, whether at the level of individuals, firms or governments, takes place in a complex environment. Many factors, some ephemeral and some more permanent, combine to influence the outcome of any particular decision. We have seen the implications of this. Intent may not lead to the desired outcome. Whether drawn up by the mice and men of Robert Burns's famous poem, by multinational corporations or by governments, even the most carefully constructed plans may lead to unexpected and even undesired outcomes.

We do not want to adopt an almost nihilistic view that nothing can be done. Agents, people, firms and governments are not completely powerless in the face of the iron curtain of uncertainty which is drawn over the future. There are chinks of light. So in the Hotelling game, for example, we can offer a framework for thinking constructively about the problem, about the assumptions we might make and the information we need to see if such postulates approximate the real-world problem we are analyzing, and we might be able to make a decision which gives us a slightly better chance of success than if we choose purely at random. In the Schelling game, in the first instance an individual *can* make the move which places him or her in a location which satisfies the individual's preferences. This may persist over time, and the person will remain happy. Macbeth was, after all, for quite a long period of time able to enjoy the fruits of his dastardly deed and become King of Scotland. But we cannot guarantee that this will happen, nor have we the ability to foresee whether it will, because the eventual outcome depends not only upon the future decisions of other individuals but upon the interactions between the consequences of these decisions.

In chess, we can draw on the knowledge and experience of others to help us make better moves. We may only understand completely a tiny fraction of all possible positions in the game, but we can devise strategies to cope with this massive level of uncertainty. In many games, and not just the wholly trivial one of noughts and crosses, we can learn strategies that avoid obvious loss. We may never be able to discover the very best, the optimal strategy – or indeed recognize it if we stumbled across it by chance – but we might find reasonable ones.

8 Doves and Hawks

Economists have reflected on the problems of trying to understand how complex systems operate. It is not that they are unaware of such difficulties, although, sadly, it is the vast majority of younger economists, emerging clutching their MScs and PhDs, who appear to have been socialized most effectively from an early stage to avoid thinking along lines which carry the slightest hint of heresy.

Some fifty years ago, Armen Alchian, whom we met at RAND in Chapter 5, drew parallels with the world of biology. He put forward what has become the classic statement on reconciling maximizing behaviour and the existence of uncertainty. Milton Friedman, writing at around the same time, arrived at a similar position. Agents, Alchian argued, are assumed on the one hand to maximize their individual utilities, yet on the other it is recognized that under conditions of uncertainty it is impossible for individual agents to follow maximizing behaviour, because no one knows with certainty the outcome of any decision. He sought to reconcile the two views, deeming that maximization nevertheless still occurred because competition dictates that the more efficient firm will survive and the inefficient ones perish.

In other words, by taking Darwin's maxim of 'survival of the fittest' quite literally, Alchian argued that the most efficient firms survived because it was as if – that quintessential economist's phrase again! – they behaved as full information maximizing agents. The process of natural selection would ensure that those agents who, consciously or otherwise, followed such a strategy would prosper.

There are indeed parallels to be drawn with biology, and during the second half of the twentieth century biologists themselves have in turn used concepts from economics, such as game theory, to good effect. Later in this chapter we will discuss examples of successful applications of these concepts by biologists to understanding aspects of evolution, and we also consider the partial and restricted nature of this success. But, first, we need to think about the 'as if' argument of Alchian and Friedman. How useful is this parallel, made fifty years ago and still invoked today, between economic and biological evolution?

On the face of it, there is an immediate problem when comparing classical economic theory with biology. We can think of two extreme postulates with which to describe the rules of behaviour followed by individual agents in any particular game or social, business or economic context. At one end of the spectrum, we have the full-information, rational maximizing agent of conventional economics. An individual equipped with these powers has the cognitive ability to gather all available relevant information and to process it efficiently to arrive at the optimal decision. At the other, we have the agent possessed of no cognitive ability whatsoever, with no information and no way to analyze it usefully. Agents such as these behave as if they are acting at random. In terms of evolution, this is exactly what does happen; it is not merely a theoretical postulate about behaviour. It is known to be true. Genes cannot mutate with intent. A species cannot decide to become fitter.

But there is a particular set of contexts in which we might agree to keep economists happy by going along with Alchian's argument. Suppose many pairs of individuals play a game of tossing a fair coin and guessing the outcome. They may be staking money on each toss of the coin or they may, in a more gentlemanly way, simply be keeping the score and seeing who wins over many repeated plays of the game. In such circumstances, it would be impossible to distinguish between the two extremes of how individuals are presumed to behave. A rational, maximizing player would collect information on the various outcomes of the tosses of coins, analyze them using sophisticated mathematics and conclude that they were random. So there is no strategy which he or she could adopt on the basis of all available information which would lead to an outcome which was anything other than random. Over a long sequence of plays, an individual might win or lose overall, but there is nothing that the person can do to affect the outcome. An agent who gathered no information at all and who adopted a very simple behavioural rule such as, say, calling 'heads' every time would on average do just as well as a rational maximizer. A chicken can play this particular game just as well as a tenured professor of economics.

If large numbers of pairs are playing the game, probability theory tells us that most players will not win or lose very much, but a few will have substantial winnings and a few substantial losses too. However, it would be positively misleading to say that the select group of winners had found a better strategy, that natural selection had chosen those who were acting as if they were full-information maximizers.

A much more realistic example of this is in the management of money by unit trusts, banks or brokers. We came across the efficient market hypothesis at the end of Chapter 6. Economists themselves have assembled a huge and impressive volume of empirical evidence proving that it is not possible to construct a strategy based upon the past behaviour of asset prices that will consistently outperform the market as a whole. No matter how much effort is expended, no matter what techniques are used, this result will hold. In other words, a strategy of choosing a portfolio at random is identical to that of a full-information, rational maximizing agent.

So it is impossible to tell in such a situation whether or not it is 'as if' the better performing agents have somehow managed to follow a maximizing strategy, albeit one which they themselves are unable to articulate consciously. It is therefore much more logical to prefer the simpler hypothesis, namely that the outcomes that we observe arise at random, which is also entirely consistent with the empirical evidence. In other words, we do not need the panoply of cognitive powers and rational information gathering and processing to understand the outcomes in financial portfolio performance.

But more generally, to say that the fitter survive in the process of competition is not at all the same thing as saying that only the very fittest possible survive, these latter being the agents that have gathered all available information and processed it to follow a maximizing strategy.

Many different hypotheses to both the hypothesis of maximization and very high levels of cognitive ability, and to the one of zero cognition and hence random behaviour, are also available. As we have seen, for example, bounded rationality has proved a useful tool in economics. Because it retains the assumption of maximization, that agents choose the best decision on the available information, it is close to the fully rational end of the spectrum of possibilities about individual behaviour. But by allowing for imperfect information, it admits that agents might not take the very best decision possible, the one they would take if they had access to all information. The survivors we observe could equally well be acting as boundedly rational rather than fully rational agents.

Close to the other extreme, we can postulate that it is as if most agents lack cognitive ability and take random decisions, but that a minority are able to discover a rule of behaviour that gives them a better than random chance of survival. Some of the majority will survive by pure chance. The rule discovered by the minority might only give them a very slight edge, like discovering that a coin is not in fact true but is biased 51:49 in favour of heads. Some of these will be made extinct at random, through bad luck, but a greater proportion of them will survive than will amongst the rest of the population. We will see in Chapter 13 that in fact there is strong empirical support for this particular hypothesis in the context of firm extinctions.

But the point here is that the outcome which we observe – at any point in time some agents have survived and others have not – does not mean that we can conclude that in some way the process of evolution has weeded out all but the very best, that it is as if only maximizers survive. What we observe at the overall level in terms of survival and failure is in principle consistent with a whole variety of hypotheses about individual behaviour. We can only distinguish between them by means of empirical evidence, by which theory is closest to the facts. But, as mentioned above, this is a key theme of the closing chapters of the book.

In any event, thinking within biology itself has moved on since Alchian and Friedman sought to draw parallels with evolution half a century ago. Implicit in their argument is a view that progress takes place, that, gradually, the less efficient will be weeded out and the most efficient will necessarily survive. The possibility that efficient firms, too, can fail is left out of the picture.

Biologists no longer claim that the process of evolution has had purpose and that a necessary outcome of such a process is for species to emerge with ever-increasing levels of fitness. Indeed, in terms of formal mathematical models of the evolutionary process, we will see in Chapters 10 and 11 that this is most definitely not the case. All we can say is: at any point in time, the species which thrive happen to be the ones which the random process of evolution has led to thrive. And, at some stage in the future, they, too, will become extinct. Even the fittest might fail.

In the case of biological evolution, it is, of course, the myriad of individual random mutations rather than the conscious decisions by individual agents which drive the system. It might seem that this makes the process too complicated to be represented meaningfully in a scientific model. Models, as we will recall from the discussion of the Schelling model in Chapter 4, are like maps: they are intended to inform us about reality. But in order to do so, they need to make dramatic simplifications.

During the course of the twentieth century, biologists succeeded in successfully illuminating certain aspects of the process of evolution, but they did so by making simplifying assumptions in their models which abstract quite dramatically from reality. The assumptions are sufficiently big as to limit the insights which these theories can give us. Nevertheless, they do tell us useful things both about how species interact with each other and how evolution gives rise to situations in which competing species live together in symbiotic harmony.

In the first instance, we will make a really strong assumption, namely that the process of evolution at the individual level is frozen. In other words, we imagine that the random mutations which lead to increases or decreases in the fitness for survival of individual creatures are temporarily suspended. This may seem at first sight to be a distinctly odd thing to do, especially in the context of a discussion of evolution itself, but it is not at all unreasonable provided that the timescale we are examining is sufficiently short. Of course, evolution is a continuous process that happens all the time, so we know that the assumption is in a literal sense wrong. But over the course of a decade, or a century even, for many creatures, particularly the larger mammals, it is not a bad approximation to make. On this timescale, the process of evolution is extremely difficult, if not impossible, to detect.

Biological species themselves, the collections of similar individuals, interact in complex ways. Different types of creatures compete for the use of the same set of resources. Herbivores, for example, may feed on grass and, if this is in limited supply, the success of one type of species is at the expense of another. Other kinds prey on these same beasts and at the same time compete amongst themselves; and the fittest survive.

The relationships between species are even more complicated than this. Sometimes, we can observe overt co-operation between different species, as with birds that are allowed to feed on the fleas and ticks nestling in the coats of dangerous wildcats. In the words of the doggerel verse, 'Bigger fleas have little fleas upon their backs to bite 'em, little fleas have lesser fleas, and so on, ad infinitum.'

There are close parallels with these interactions, at the same time both competitive and co-operative, in the world of business. Firms obviously compete with each other, whether directly in the marketplace for the goods and services which they produce or indirectly for resources such as labour. Almost all economic theory is concerned to describe this competitive aspect of behaviour. Even more widespread, however, is co-operation between firms. We are not thinking here of collusion between firms in terms of, say, setting prices, an activity that has been illegal in most western countries for many decades. We saw in Chapter 1 how firms in America reacted to the dramatic upsurge in competitive pressure towards the end of the nineteenth century by forming trade associations and cartels, and how the federal authorities legislated to strike these practices down.

The basic point here is that the output produced by most firms is bought not by individual consumers but by other firms. Even the smallest firm buys in supplies from other firms which are an essential part of its ability to produce its own product. Tyre manufacturers, for example, sell to the large motor companies, so when General Motors does well, their suppliers benefit. There are other aspects of co-operation. For example, in high-technology industries, firms may agree an industry standard for component parts which they all need.

In other words, the economy can be thought of as a system in which each of the individual component parts is connected to others, with flows along the connections. Positive energy flows across many of these connections, for the greater the success of any particular firm, the more prosperous will become lots of firms with which it has dealings. Some connections will carry negative energy, where the firms are competing and where the success of one is at the expense of the other, but the economy is best conceptualized, best analyzed, as a connected system. And the best way of understanding its behaviour is to understand the pattern of the connections and how they behave. Economic theory, in contrast, not only ignores many of the connections that exist in practice but also makes artificial segmentations of the system into categories such as 'the labour market' and 'the goods market'. We need to focus on the system in its entirety.

In biology, there is in addition a much more subtle form of co-operation, which is observed very widely. Species that at one level are in mortal combat nevertheless depend upon each other for their very survival. We see this occasionally in the economy, but much less frequently. The most commonly observed example is in the markets for new products, when consumers require education and persuasion not just to buy a particular brand but to convince them that they should buy the product at all. The success of one firm can help spread awareness of the product and so assist its rivals.

Picture the scene at a watering hole in the African bush, vital for survival in the arid, sun-drenched landscape. All kinds of creatures gather to drink, from fragile antelopes to huge, powerful water buffalo. Each of them has had to adapt, to evolve a strategy that enables them to take their place around the water. Further afield the predators lurk: large cats, waiting patiently to devour a tasty meal. These, too, have evolved to find a niche that enables them to survive. Their stealth, pace and strength enables them to overpower their chosen prey. But the victims, too, have evolved, their acute sensitivity and their agility giving them some chance of avoiding the teeth and claws of the predators.

Here is a snapshot of evolution, taken at a single moment in time, showing the complex relationship between the different kinds of predator and prey species and the competition within the two groups. A giant cat which was *too* successful in its ability to catch its prey, whose appetite was *too* voracious would soon become extinct itself as the population on which it fed became destroyed. Likewise, a species endowed with the clumsy ability of a Tom to catch his Jerry would also soon disappear. Amongst the prey, a species which lacked the ability to find a niche at the watering hole, a time when other creatures did not crowd it out from using this scarce and vital resource, would wither and die.

The same picture can be seen wherever we look in the natural world. A famous – famous, that is, within the abstruse mathematical world of statistical theory – series of data records the number of lynx trapped each year between 1821 and 1934 at the Hudson Bay trading post (see Figure 8.1). The series fluctuates dramatically from year to year, and a challenge for statisticians is to find a way of discovering any underlying regularities in the way the numbers move. Biologically, too, the series is famous. Lynx prey upon Arctic hares. Some years, they are particularly vigilant and successful in their task, and the number of hares declines. But then starvation overtakes the unfortunate lynx population because their prey has become scarce, and the lynx population declines. This in turn gives the decimated hares the chance to breed and multiply. And so the process goes on through cycle after cycle.

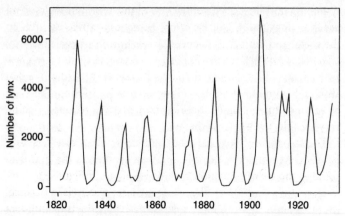

FIGURE 8.1 Number of lynx trapped at the Hudson Bay trading post, 1821–1934 (Source: *Non-Linear Time Series Analysis: A Dynamic Systems Approach*, H. Tong, Oxford University Press, 1990).

The complicated relationship between individual lynx and hares enables both *species* to survive. They are mortal enemies at the individual level, yet as species they are bound together as if they were in co-operation. Individual numbers fluctuate wildly, yet both species continue to exist. It seems that evolution has brought the lynx to the point where they are very good, but not *too* good, at catching hares, while hares are skilled, but not *too* skilled, at avoiding the fate of being the dish of the day. But each without the other would be in danger of becoming extinct. Lynx need the hares to eat, while, without their natural predator, the hare population would grow without constraint, and as a consequence they would run the risk of eating so much of their own food sources that these, in turn, fell below the minimum level of population needed to sustain and reproduce themselves successfully.

This type of relationship between species was formalized in the 1920s by the great Italian mathematician Vito Volterra. He used it to explain, amongst other things, the movements in fish population in the Atlantic, and he expounded his theoretical and empirical findings in a book, *A Mathematical Theory of the Struggle for Life*. The particular mathematical structure invented by Volterra, a system of non-linear differential equations, is still in widespread use in biology today. The use of mathematics is not as essential in the social and biological sciences as it is, say, in physics. Darwin, after all, wrote entirely in English, and many of the truly great economists – Smith, Ricardo, Marshall, Keynes – relied only in an absolutely minimal sense, if at all, on formal maths. But maths can be very helpful in making theories more precise and in producing very specific, testable hypotheses.

The struggle for survival as described by Volterra offers us a paradox. At one level, there is constant change, as the dramatic shifts in the lynx population over time shown in Figure 8.1 illustrate. Yet, in a deeper way, the whole intricate system is in balance, in equilibrium. The random mutations of individuals that underpin the process of natural selection, of survival of the fittest, have brought the lynx and hares to the point at which they co-exist. In conflict at the level of the individual, harmony prevails at the level of the species. This is but a single illustration of a phenomenon which is seen everywhere in the natural world: species co-existing whilst at the same time being in intense competition.

Volterra's influential work can be thought of as a very early formalization of a part of the theory of evolution. It is very partial, for it does not deal with how species change and evolve over time. Rather, the theory describes clearly a subtle way in which species might be connected. It gives a snapshot of the outcome of the process of evolution at a point in time. But we can also use it in the case of a particular example to ask what changes would be required in order to bring about the extinction of the species locked together in this symbiosis. The lynx and the hares have evolved to the point where each is essential to the survival of the other. We might be able, with skill and judgement, to write down a particular example of Volterra's theoretical equations, just as he did with fish populations, which will describe the observed movements over time in the two populations. We then ask: what if? What if a change in the external environment – a particularly severe winter, for example – kills off too many lynx or hares for the species to survive, or what if either of them gradually evolve over the course of generations to become more speedy? How far can the particular circumstances be altered before the delicate and intricate balance between the two species is destroyed?

Questions such as these are not merely of purely intellectual interest. The Volterra approach and these 'what if?' issues are matters of everyday practical concern in understanding, for example, the depopulation of certain species of fish and how policy might be designed to deal with the problem.

We can also go on to ask: how do situations such as this arise? What is the process by which this dependency between competing species can come about? To do this, we need to relax the assumption of the discussion so far that the process of evolution is frozen. We introduce evolution back into the model explicitly. And we do so in the context of game theory.

Biologists have made more successful use of game theory than economists. Imagine, for example, a population in which the individual members have two strategies open to them which guide their everyday behaviour. Hawks, as we shall call them, are fiercely aggressive and always fight to injure or even kill their opponents, but this strategy is not without risk. The act of aggression itself creates the possibility that the perpetrator itself might be damaged in the process. In contrast, the Doves, whilst they may flourish their tails or feathers, are careful always to avoid conflict unless they are actually attacked. Of course, we are not discussing in a literal sense any actual population of doves and hawks, but merely using these as labels to describe different sorts of behaviour within any particular population.

A population of Doves leads a life of Elysian calm amongst the trees or prairies, depending upon whatever kind of species we might think of it as being. Unfortunately, their state of primordial bliss is open to attack by mutant genes within the Dove population. The random process of mutation gives rise to an individual with a much more Hawk-like attitude. This fiercer beast has many opportunities for success amongst the docile population, and its genes may spread. The Dove population is susceptible to invasion by the mutant. At the opposite extreme, a population comprised entirely of aggressive Hawks would be stressful and costly. Every time a potential conflict arose, the only strategy available is one of attack. Here, a mutant could prosper by the simple expedient of avoiding conflict.

An illustration of these principles can be found in the Evelyn Waugh novel *Decline and Fall*. The hero, though blameless, through a series of unfortunate accidents finds himself expelled from the University of Oxford, where he led a modest life and studied theology. Obliged to work for a living, the only thing he is remotely qualified to do is teach. Given the circumstances of his expulsion, he finds himself teaching in the lowest circle of the English independent boarding-school system. Here, he encounters a true monster, his fellow schoolmaster Captain Grimes. Grimes's whole career is based on following strategies which are so deviant from those of the societies of which he is a member that he thrives and prospers. An officer in the Army, he is sentenced to death in Flanders during the First World War for cowardice in the face of the enemy. Left alone with a bottle of brandy and a revolver to do the decent thing, he simply drinks the brandy. His colleagues are so embarrassed that they engineer not his execution but his release. A serial paedophile, his behaviour is so open and outrageous that headmasters seek to cover it up and merely dismiss him from his post, with glowing references.*

Clearly, Grimes follows behavioural rules that are completely contrary to those which are deemed normal and reasonable, in whatever circumstances he finds himself. But if other people started to follow his strategies, then, no matter what the difficulties, some of those following a Grimes strategy would surely have been shot or handed over to the police.

In the games we considered in the previous chapter, the rewards available from each particular strategy were fixed. But suppose that they are allowed to vary over time. There are many ways in which variations can take place, which serve merely to make the

* I am assured that this practice is now far less prevalent in the English independent-school system than it was in the first half of the twentieth century, when Waugh wrote his novel.

analysis of the game even more complicated but do not lead to greater enlightenment about which strategies are best to follow. There is one specific way, however, which enables us to tell a plausible story about how a mixed population of Doves and Hawks can not only emerge but prosper. A solitary Hawk has a great advantage in every single conflict in a population otherwise made up of Doves but, as its successful genes spread and aggressive behaviour becomes more common, the rewards from ferocity gradually diminish. Increasingly, Hawks encounter like-minded individuals, and a strategy of aggression gives rise to declining levels of reward. Eventually, we may readily imagine that a kind of equilibrium will emerge in which the potential rewards from playing either strategy are equalized and Doves and Hawks co-exist in a kind of symbiotic harmony.

However, it is another example from English literature which begins to convey the limitations of this kind of analysis. H. G. Wells's short story *The Country of the Blind* tells of an explorer who survives harrowing experiences in the South American Andes and finds an oasis of civilization in a virtually impenetrable valley. But this society has a curious feature: all its inhabitants are blind. Once recovered from his ordeal, he attempts to base his strategy on the old adage, 'In the country of the blind, the one-eyed man is King.' However, far from guaranteeing mastery over his new-found neighbours, his sojourn ends in ignominy. Rather than becoming king, he eventually takes the decision to flee the country across the dangerous mountain terrain.

Here, a population resists invasion, as it were, by an apparently better equipped and potentially rival type employing a different strategy. The sighted hero has undoubted advantages in any single encounter with a member of the blind indigenous population, but they in turn have evolved complex strategies to enable them to deal with their lack of sight. For example, they choose, perfectly sensibly, to work in the cool and sleep in the warm. Yet, for the would-be monarch, this involves the not inconsiderable inconvenience of working during the night and trying to sleep during the day. He attempts to convince them of his superior powers and to explain what sight involves. He claims to have observed individuals at certain parts of the village, for it is laid out on a regular grid basis, at certain times. But, having made himself unpopular with some, when questioned the relevant people deny their whereabouts. The hero, or rather anti-hero as we would see him nowadays, goes on to attempt predictions and announces when a certain person will arrive at the spot where his audience is. Equipped with superior hearing, the individual concerned deviates from his apparent intent, and the forecast is falsified. Even more complications are introduced by the fact that the putative king falls in love with one of his potential subjects and his resolve to rule is weakened.

Wells's story shows the limitation of game theory, even in its evolutionary form when the rewards can vary over time, once the situation becomes at all complicated. If we observe two species, or two varieties of the same species, co-existing in a kind of harmony, simple game theory can be used to describe how this situation has come about. Initially, in our Doves and Hawks example, a more ferocious mutant Dove is able to invade the population, because initially the rewards for this Hawk-like behaviour are high. The mutant has discovered a superior strategy to play. But this gives no guarantee at all that a superior strategy will *necessarily* lead to a successful invasion of an existing population. The problem once again seems to stem from the complexity, from the dimension of the problem. If it is reasonable to make huge simplifications when describing the environment in which different strategies exist and compete, then game theory can be very helpful. But once we introduce more than a few key factors into a situation, it becomes much less useful for understanding what is going on.

One way of interpreting this, of course, is to say that the level of uncertainty is high precisely because so many different balls need to be kept in the air at once. The central character in Wells's story believes initially that he is playing a straightforward game in which he will be able to invade the population, but it soon becomes clear that the problem is many-layered, is multi-dimensional.

In any event, game theory says nothing to us directly about the central empirical features of failure and of extinction. The pace of extinction appears to vary over time in irregular waves, but there is a very definite relationship between the size of an extinction event – what percentage of species become extinct in any given time interval – and how frequently we observe it. It is to this vital evidence that we now turn.

9 Patterns in the Dark

Since the beginnings of life on earth aeons ago, almost all of the millions upon millions of different species which have tried to adapt to the environment, to carve a niche for themselves in which to survive, have failed. And they have failed in a particularly dramatic manner. They have simply ceased to exist. From the terrifying dominance of *Tyrannosaurus rex* and the other flesh-devouring dinosaurs to the most humble mollusc or earthworm, almost all have failed. Failure is by far the single most important feature that biological species have in common.

Our understanding of why this has happened was advanced in spectacular fashion by Charles Darwin's brilliant theory of evolution, expounded in the middle of the nineteenth century. Darwin's ideas on evolution may appear commonplace today, but at the time they were revolutionary. His genius consisted not just in formulating the theory of evolution but in providing empirical evidence to support his theory. He was not an ivory-tower scientist, theorizing about what the world ought to be like. He tried to test his ideas against what the world is actually like. A hundred years before him, Adam Smith, one of the greatest economists who has ever lived, illustrated his theories in *The Wealth of Nations* with immense detail from the whole panoply of recorded human history. Darwin, too, in his masterpiece *On the Origin of Species*, carefully sought out examples from all over the world, from all times and places, with which to test his theory.

This juxtaposition of the political economist Adam Smith and the biologist Charles Darwin is not accidental. This book is about the striking parallels between evolution and extinction in the biological world and in the sphere of human activities. But, first, we must examine what Darwin's theory tells us about extinction, about failure, and we must then set this in the context of some very recent and important developments, both empirical and theoretical, in the field of evolution.

Darwin's theory is usually regarded as being about survival rather than extinction. 'Survival of the fittest' is no longer confined to a specialist, biological context but has become a phrase in general use within the English language. The theory of evolution embraces the complex relationships between different types of species and how each of them changes, or evolves, over time. Evolution takes place at the level of the individual being, from the humblest bacteria to the incredibly sophisticated *Homo sapiens*. No single offspring is a complete carbon copy of its parents. Subtle differences, sometimes barely detectable at the level of the species for many generations, arise from the random mutation of genes which make up the individual. On occasions, we can witness dramatic transformations. A principal reason why AIDS is so difficult to cure, for example, is that the HIV virus itself mutates in a remarkable fashion. The speed and scale of change is breathtaking.

There is a large literature in biology which analyzes the process of evolution at the very basic level of the gene. As early as 1927, for example, the British biologist J. B. S. Haldane was developing mathematical formulae to help model the processes involved and the effects of reproduction and selection on the fitness of a population. A Scot educated at Eton and Oxford, his major work, *The Causes of Evolution*, placed natural selection on an even firmer scientific basis by explaining it in terms of the mathematical consequences of existing theories of genetics. He was responsible for many very quotable aphorisms, of which perhaps the best is his description of the various stages through which a new scientific theory passes: 'There are four stages of acceptance: i) this is worthless nonsense; ii) this is an interesting, but perverse, point of view; iii) this is true, but quite unimportant; iv) I always said so.'

Essentially, there are two types of reproduction that occur: asexual and sexual. Asexual reproduction is when an organism makes a direct copy of its genetic material and then splits into two, creating two organisms identical to the parent in terms of genetic material.

Sexual reproduction involves two parent organisms which mate and produce offspring whose genetic material contains genetic material from the parents' genomes.

Interest in the conditions under which sexual reproduction offers advantages over asexual reproduction goes back considerably further even than the work of Haldane, perhaps to the German biologist August Weismann in the 1880s. It might be thought obvious that sexual reproduction confers an advantage compared to asexual, since there is a chance that the offspring will have higher fitness than the parents, whereas with asexual reproduction it is the same.* Nevertheless, despite this long history of scientific enquiry, it cannot yet be said that the process of evolution at this level is understood completely and, since this book is primarily about social and economic issues rather than biology as such, it would be a long diversion to describe the development of this work.

For our purposes, however, a very important finding does permeate this literature. Namely, that the probability of a gene successfully invading a population – 'fixing' in the jargon of biology – is on average very low. This is the case even when a mutation takes place in a population which reproduces asexually, and the mutation reproduces sexually. The mutation itself might fail to 'fix', for example, because it has a low overall level of fitness, so its chances of spreading through the population are themselves low. Even a high level of fitness is not in itself a guarantee of success, for it might occur in a population which happens to be at an even higher fitness level. So most mutations, most innovations, fail.

These changes at the level of the gene take place at random. Those which tend to increase the chances of survival have a greater chance of persisting, of being passed on to future generations, than those which do not. At the level of the organism, we can observe successful examples. Sunflowers, for example, have gradually evolved the ability to turn their faces to the sun as it moves across the sky. Those that can do this are more likely to

* Except for mutations that occur during the copying process.

survive than those that lack this facility. The fittest survive. However, as the outstanding biologist and writer Stephen Jay Gould observed, an even better strategy for the plants would be to grow feet, so they could follow the light still more effectively. Yet it is not within the capacity of sunflowers, or indeed of any creature, to plan to grow feet, to act with the conscious intention of producing a mutated variant of the species possessing feet. The changes which gave the plants the ability to turn their heads and follow the sun took place at random.

The fact that evolution takes place at random might suggest that we cannot discern any patterns in its history. A commonplace example of a series of random data is that generated by successive shakes of a dice. Provided that the dice is true – in other words it is not biased in any way – the outcome we observe will be a random string of the numbers between 1 and 6. There will be no apparent pattern. We cannot use information about what happened on previous shakes to predict successfully what will happen on the next one. No matter what the previous history has been, on the next shake each one of the numbers from 1 to 6 is equally likely to appear. Even if, say, there have been six successive shakes of a 6, which simple probability theory tells us has about a 1 in 50,000 chance of happening, on the next shake a 6 has the same probability of coming up as each of the other numbers.

This concept, though proved beyond doubt, is often intuitively hard to grasp. An example is the various national lotteries which take place around the world. A set of numbers is drawn at random, and the lucky person holding the ticket containing these numbers makes a fortune. A small industry has developed of people purporting to be able to use information from the past sequence of winning numbers to increase the chances of winning this time. Sometimes they will simply supply the information on past numbers for the client to manipulate to his or her own satisfaction. But these products have no value at all in the sense of enabling successful predictions to be made. Those who buy them do so in ignorance of the fact that lottery numbers are drawn purely at random. Of course, they do have value in a different sense in that people are prepared to pay a price for them.*

Yet there *is* a very distinct pattern we can observe in shakes of a true dice or sequences of winning lottery numbers, both the outcomes of random processes: given a sufficient number of shakes or draws, we will in general see that each of the possible numbers occurs a very similar number of times. In the case of the dice, the number 1 will be shaken almost exactly one sixth of the time, as will number 2, and so on.

So, from a random string of numbers, we can discern order and pattern. There is a very clear structure to the probability distribution, to use a technical phrase, of the outcome which we see. The probability distribution tells us what proportion of times each of the individual numbers will be observed, given a sufficiently large number of shakes or draws. This is determined very clearly and decisively.

Very recent work analyzing the fossil record of extinctions of biological species has performed the same sort of task. Yes, evolution does take place at random, but we now believe we can discern patterns in the distribution of probabilities of the outcomes of this random process as far as extinctions are concerned. The analysis is rather more subtle than the simple analysis required to observe patterns and structure in the outcomes of shaking dice or drawing lottery numbers, so it requires a little more patience to follow its logic. But the evidence is both clear and dramatic.

One thing we now understand is that evolution seen from the perspective of the system as a whole, of the total number of species, offers us an intermingling of stability and change, of equilibrium and apparent disorder. The juxtaposition of these two concepts

---

* Some might see here a rather chilling real-life parallel with George Orwell's novel *1984*. Set in an imaginary future of an England under totalitarian communist rule, the Party treats the ordinary workers, the Proles, with utter contempt and keeps them in poverty. One of their few recreations is the weekly lottery. Orwell's hero, Winston Smith, observes that the Proles, whilst lacking all but the most elementary education, are capable of stupendous intellectual feats of recalling past sequences of winning numbers. In the pubs, they engage each other in fierce argument about the numbers most likely to appear in the next draw.

is the key way in which our understanding of Darwin's original insights has developed – evolved, one might even say – in the closing decades of the twentieth century. And the advances have been both empirical and theoretical.

One of the most striking developments that has been made is the realization that the pace of evolution varies widely over time. Periods of relatively slow, steady changes appear to be interspersed with periods of dramatic change. Stephen Jay Gould formalized this concept in the phrase 'punctuated equilibrium'. As we mentioned at the beginning of Chapter 1, his outstanding book *Wonderful Life* describes the proliferation of different types of life which emerged in the early Cambrian era, some 550 million years ago. For aeons, the pace of evolution had been almost imperceptibly slow. More advanced forms had emerged from the basic single-celled organisms with which life began, but the time span over which this happened is measured in thousands rather than hundreds of millions of years. Yet, suddenly, in the biological equivalent of the twinkling of an eye, many new forms of life suddenly emerged. And most of these became extinct. They failed.

We can see the mirror image of the speed of change of evolution in the fossil record on species extinctions. This has slowly been pieced together by immensely careful and painstaking research. The key figure in this was the colourful University of Chicago palaeontologist J. John ('Jack') Sepkoski Jr. He documented exhaustively the ups and downs of life through the last 600 million years. By collecting the data and developing a series of statistical methods to study it, he gave us a new way of understanding the history of life.

With his Chicago colleague David Raup, Sepkoski developed hypotheses which were startlingly original and controversial. Perhaps this is not surprising, given his unconventional taste in music: his favourite group was the punk-rock band The Sex Pistols, and he considered the Velvet Underground and Nico album to be the greatest ever made.

One of the particular problems of analyzing the fossil record in a statistical way is that there is no single, universally accepted set of data. Sepkoski worked on his database for close to the whole of his professional life and the results he published refer to various versions of his database. Moreover, he provided copies of his database to others at various stages of its development, and these are being traded now on an informal basis. The canonical version of species extinction simply does not exist.

For example, Raup's initial work was based on the evidence on extinction of 2,316 marine animal families in the seventy-nine generally recognized stages of geological time. He later refined the analysis with a much larger sample of almost 20,000 genera, but the problem of determining the appropriate set of data to use is even more complicated. There are series for extinctions across all species, series for continental life, series for marine life, each of which have their own strengths and weaknesses. To make matters even more complicated, there are minimum and maximum estimates, relating to the degree of certainty which can be attached to the appearance of the first and last records of a species in the fossil record.

Economic data is subject to a great deal of revision and uncertainty. For the distant past, by which we mean in this context not 500 million years ago but the first half of the nineteenth century, controversy still exists about what actually happened. However, for the most part, the data series which are analyzed by economists, though subject to many caveats and uncertainties, are usually agreed across the profession. Everyone analyzing, say, American output growth since 1950 will use the same set of data. Even this little luxury is denied to palaeontologists.

The differences between the numerous data sets should not be exaggerated, for the sets contain many similarities. Everyone, for example, has heard of the extinction of the dinosaurs some 65 million years ago, but there are other events in the fossil record which indicate extinctions on an even bigger scale. Some 250 million years ago, for example, over 90 per cent of all marine species became extinct very rapidly. Biologists as long ago as the nineteenth century realized that there had been five very large extinction events, which they used to define boundaries in the biological timescale over the past 550 million years. Each of the periods between these events is marked by the dominance of quite different kinds of species than in the others. The names of these epochs may be dimly familiar from the classroom: Ordivician, Devonian, Permian, Triassic and Cretaceous. There were other, less pronounced but still distinct periods of substantial extinction, and these, too, were identified and named as sub-periods of the 'Big Five' periods.

The most contested theory put forward by Raup and Sepkoski is that catastrophic extinctions of species may have occurred approximately every 26 million years during the past 250 million years of the earth's history. These events also included the extinction of the dinosaurs 65 million years ago. In other words, large-scale extinction is not merely a random event but has a pattern in terms of how frequently we observe it. The period over which the event occurs is constant.

The theory has attracted attention from a wide range of scientific disciplines. It has generated cult beliefs, such as the idea that the sun has an unobserved stellar companion, orbiting at a distance of about one light year. Every 26 million years, the orbits of the sun and Nemesis, the sinister name given to this putative companion, interact in a way that brings a massive shower of comets into the solar system, bombarding earth with terrifying impacts and wreaking destruction on a planetary scale.

The problem with what is often called the 'Nemesis' hypothesis is only partly one of differences between different sets of data. More fundamentally, the evidence is not quite strong enough for it to be established, even when analyzed using powerful statistical techniques the details of which need not detain us here. There is just enough in a reading of the runes – or an analysis of the data, to give the activity its more scientific name – to convince a true believer, so the hypothesis cannot be discarded completely, but the evidence is weak.

In fact, the first impression that a plot of the extinction record over time gives is one not of regularity but one of marked irregularity, of a lack of clear periods in the patterns of fluctuations. We could plot the percentage of all species becoming extinct in different geological periods over the past 500 million years or so, with the period as a whole divided into its seventy-seven sub-periods.* However, an important problem with the data in its raw form is that the time lengths of the different geological eras represented in the chart themselves differ. The shortest lasted 2.5 million years and the longest 8.5 million, so we would expect some quite large differences in the percentage of species becoming extinct simply because the longer the period, the higher we might expect extinction to be.

We can get round this problem easily by measuring the percentage of all species thought to have become extinct within a particular period of time; one million years, say. This way, we can compare different periods. By taking the percentage of species becoming extinct in each case, two periods can be compared in terms of the scale of extinction. We are measuring in each case the percentage of species that becomes extinct every million years. Figure 9.1 plots the data in this way.

There is an apparent downward trend but otherwise at first sight

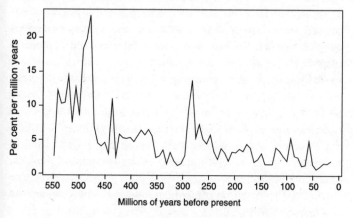

FIGURE 9.1 Percentage of species becoming extinct per million years.

* This was very kindly provided to me by Professor Norman Macleod of the Natural History Museum, London, and is the Sepkoski 1998 database, which is the last one provided to colleagues before his premature death from a heart attack in 1999.

the series appears to be irregular. Extinctions were particularly high around 500 million years ago, following the explosion of different life forms which we noted earlier, but there are no obvious regularities in the data, no obvious patterns indicating that large extinction rates follow each other at regular intervals of so many millions of years.

We should say immediately that a thorough analysis of patterns of regularity in data such as that plotted in Figure 9.1 involves far more than the simple inspection of the data in a chart like this. Even when using the most sophisticated mathematical techniques, plotting the data is a simple but nevertheless essential first step to take. But the more technical analysis does come up with a glimmer of support, but no more, for the Nemesis hypothesis, even though it is not at all obvious in Figure 9.1.

The initial impression of the plot of data in Figure 9.1 is that it has the same property as the data generated in a series of throws of a dice or in the drawing of lottery numbers. Plots of extinctions or throws of a dice, say, both show the history of the data but in each case with no apparent pattern. As it happens, there is more of a pattern to the extinction data. The pattern is not very well defined and is not at all obvious from simple inspection of the chart. Indeed, it requires some technical statistical analysis to identify it.* Essentially, a period with a high extinction rate is more likely to be followed immediately by one with a high rate as well, and periods with low rates are more likely also to be followed by a low rate. This is not true with a dice. Shaking a 5 or a 6 does not mean that the next shake is more likely to be a 5 or 6, or even a 4. However, when we look at the *change* in the extinction rate between periods, the resulting series is just as random as the dice or the lottery numbers.

The exciting feature of the work of Raup and others is that there *is* a further discernible pattern in the data plotted in Figure 9.1, which has important implications. The detection of this involved

---

* In technical terms, the autocorrelation function has positive and significant low order values.

some particularly subtle analysis. Raup discovered that there is a particular mathematical relationship between what we can call the frequency of the extinction rate and the size of the extinction rate.

The size of the extinction rate is easy to define. It means what it says on the label; in other words, the percentage of species becoming extinct per million years in each geological era. The frequency means how often we observe any particular percentage. In a more everyday context, we might toss a coin ten times and obtain six heads, say, and four tails. The frequency with which heads is observed is six, and of tails four.

Raup showed that the frequency with which we see extinctions of any given scale falls away with the square of the size. In other words, for example, an extinction period in which twice the percentage of species per million years become extinct as in another event is four times less likely to be observed (four being two squared). A period in which three times as many become extinct is nine times less likely to take place, and so on.

This particular type of relationship can be found in many different and diverse contexts. Handily, it has its own name, 'power law', and this is the short-hand phrase which will be used in the rest of the chapter.* More examples of this relationship, and their practical importance in business and economics, will be seen in the next chapter.

Before the wrath of palaeontologists and other distinguished researchers from other disciplines descends upon me, I must say straightaway that this empirical finding is not established to the complete satisfaction of all. There are alternative views about how the relationship between size and frequency should be represented, though the differences between the views are somewhat esoteric. Everyone agrees that a power law gives a good description of the pattern in the data, and the debate is about whether or not a slightly different type of law gives a slightly better description.

---

* The word 'power' in this context does not refer to strength or might but has a specialized usage. In maths, the phrase 'two squared' (i.e. two multiplied by itself) can be written as 'two to the power of two'. Two cubed (two times two times two) is 'two to the power three'.

The power law hypothesis is established far more firmly than the Nemesis view of the world, which postulates regular large extinction events every 26 million years. The arguments are essentially about the nuances of how frequently we can expect to see really large extinction events.

A picture can easily be worth a thousand words in conveying an idea. In this instance, however, a number of words are needed to describe what is being shown. Figure 9.2 below gives an illustration of the power law connection between frequency and size using the same Sepkoski data which are plotted over time in Figure 9.1.

There is an inherent problem, already acknowledged, about the uncertainty of the data. We cannot say with any degree of confidence that an extinction rate estimated from the fossil record of, say, 1.47 per cent of total species per million years in one particular era is less than one of 1.52 per cent observed in a different era. Obviously, if we were completely certain about the data, then 1.47 would be unequivocally less than 1.52. But there is a range of uncertainty around any particular estimate of an extinction rate. It is our best estimate of what happened, but we cannot be certain that it is absolutely correct. It therefore makes sense to get round this problem by allocating the various sizes of extinctions which have been observed into groups. By doing this, much less emphasis is placed on very small differences between individual observations.

A simple way of grouping this particular data is in bands of three, for example, so that we count up the number of times an extinction rate of less than 3 per cent was observed, how many times one of between 3 and 6 per cent, and so on up to between 21 and 24 per cent, a band that contains the highest value of extinction per million years in the data.

Of the total of seventy-seven observations, each corresponding to a particular geological era, no fewer than forty-two of them experience an extinction rate of less than 3 per cent per million years. So the frequency here is forty-two. We see an extinction rate of between zero and 3 per cent forty-two times in the fossil record.

The second step involves a small piece of maths. The bigger the

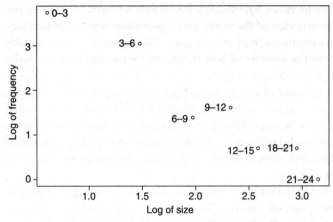

FIGURE 9.2 A plot of the (natural) log of the frequency of extinction events in the fossil record against the (natural) log of the size of the extinction. Size is defined as the percentage of all species becoming extinct per million years, and this data is grouped into bands of three, i.e. all periods when the extinction rate was between 0 and 3 are in the '0–3' observation, and so on.

size of the extinction rate, the less frequently it happens. To be more precise, as discussed above, the frequency falls away with the square of the size. This means that the logarithm of the frequency falls in simple proportion to the logarithm of the size. To be exact, it falls away twice as quickly. At this stage, the readers of this book can probably be divided into two groups: those for whom the previous statement is obvious and those for whom it is not. As always, this Manichean division of the world into two will not be completely accurate, for there are always those at the borderline between the categories, but rest assured that the statement is true.

We see at the top left-hand corner of the chart the point labelled '0–3'. This refers to the percentage of total species becoming extinct every million years in the geological eras of the data set, and refers specifically to percentages between 0 and 3. Similarly, '3–6' refers to values between 3 and 6 per cent, and so on. Reading across to the left-hand axis tells us how many times values in each of these ranges were observed in the fossil record (more exactly, it tells us the natural logarithm of this value). Reading down to the bottom axis tells us the average size of the extinction event (again the natural log) in each of the groups.

The striking feature of the chart is that the frequency with which different rates of extinction have occurred falls away as the size of the rate increases. For the point marked '0–3', we read across to a high frequency on the left-hand axis and a low size on the bottom axis. In contrast, the point '21–24' at the bottom, referring to periods when between 21 and 24 per cent of all species became extinct per million years, has a very low frequency, but a large value of its size. In other words, the hypothesis of the connection between frequency and size is consistent with the data. The number of times an extinction of any given size is observed is reduced as the size increases.

This discovery of the empirical relationship between the size and frequency of extinctions is a major advance in how we need to think about evolution. Any modern development of Darwin's brilliant theory must be able to explain it. Rival accounts have indeed been developed, as we shall see in Chapter 11. But before exploring these and their implications, we first turn to the modern economic world. Excitingly, power law, or very near power law, relationships have been identified very recently in many areas of economic activity. Perhaps most exciting of all, the relationship that describes the pattern of extinctions amongst firms appears to be virtually identical to that which describes biological extinctions. For certain types of system, as diverse as those in which biological species and modern firms flourish and die, we may have the first inklings of a general theory not of evolution but of extinction.

10 The Powers that Be

An intellectual revolution is taking place in many aspects of the biological and natural sciences, which is now being extended into the sphere of social and economic analysis. In the first part of the twentieth century, quantum physics overturned conventional views of causality at the highly micro, subatomic level of behaviour. Now, the discovery of so-called power law behaviour in widely different areas challenges perceptions of causality at the system-wide, macro level.

Very recent research shows that social and economic systems can also be characterized by such behaviour. Gene Stanley, Professor of Physics at Boston University and editor of the major statistical physics journal *Physica A*, has been a leading pioneer. One of his first discoveries in this area was that the size distribution of the largest American companies was well described by a power law, a finding subsequently generalized across all US firms by Rob Axtell of the Brookings Institution. Further, the variability of growth rates not just of companies but of whole economies appears to follow a power law relationship.

These latter may seem obscure examples. Far more relevant to most people's lives is the discovery that the structure of contacts on the World Wide Web can be described by a power law: a few sites have huge numbers of links to them, and most have very few. Perhaps even more relevant is the finding that the number of sexual contacts is also approximated by a power law: a small number of people have a tremendous number, a somewhat greater number have somewhat fewer but nevertheless lots, and so on, until we see that the category containing the largest number of people is the one where the number of sexual contacts are the least.*

Both common sense and conventional science suggest that big events should have big causes. When an enormous earthquake occurs, we should look for a major dislocation as its cause. When mass extinctions of species are found in the fossil record, we should seek a major shock such as an asteroid impact. When we see a stock-market crash, we should expect a major political or economic disaster to have happened.

Sometimes, we can find these big events. Palaeontologists generally agree that the disappearance of the dinosaurs *was* caused by such an event, even though its exact nature remains a matter of dispute. But the reasons for many of the sizeable extinction events plotted in Figures 9.1 and 9.2 remain obscure. It is clear that dramatic, large-scale events are far less frequent than small ones. In systems characterized by power-law behaviour, however, they can occur at any time and for no particular reason.

The world is not turned completely upside down by these discoveries. Most of the time, small events, small shocks to the system, will only have small impacts, and large shocks will usually have big consequences. But the fact that we observe power-law behaviour in a system tells us that the system operates in ways that mean that these relationships do not always hold. Sometimes, a very small event can have profound consequences, and occasionally a big shock can be contained and be of little import.

It is not the power law itself which gives rise to these unexpected features of causality; rather, it is the fact that we observe a power law in a system which tells us that causality within it behaves in this way. The conventional way of thinking, which postulates a link between the size of an event and its consequences, is broken.

* To be strictly accurate about a number of these relationships, power laws do give very good descriptions of the pattern which can be found. In some instances, as with biological extinctions, there is esoteric debate about whether slight variations on them, such as exponential truncation of power law behaviour, provide slightly better descriptions.

Power-law behaviour at the level of the system as a whole arises from the ways in which the component parts of the system are connected. In Chapter 8, we introduced the idea that we can think of the economy as being characterized by its network of connections; of how, for example, firms are linked to each other, and whether the flow across any particular connection is positive or negative. Sometimes, it is the distribution of the connections themselves which follows a power law, as is the case, say, with sexual contacts. Sometimes, it is the connections which give rise to power-law behaviour of another feature of the system, such as the relationship between the size and frequency of extinction events in the fossil record.

Systems which are connected in these ways are extremely difficult to insulate against shocks spreading across most or the whole of the system. For many years now, for example, epidemiologists have understood how viruses such as the common cold spread across populations. The structure of the connections, of the social network of contacts across which the common cold spreads is to a large extent purely random. Obviously, there are some patterns, such as when a child, say, catches a cold, there is a distinct chance that one or both of the parents will as well. But much of the spread of the cold virus follows from random contacts: the person sitting opposite on the train, the person in the snack bar at lunchtime, whom we may never see again. Instead of a virus, we can, of course, think of the more general concept of a shock. This could be, say, a piece of bad news which makes a firm or a trader in the financial markets pessimistic rather than optimistic; or it could be an idea, such as the view that a particular government has been in office too long and needs to be voted out.

The mathematics of how viruses – or shocks or ideas, say – percolate across random networks, networks in which the connections between the component parts – who is linked to whom – are random, has been well known for many years. Just as importantly, the principle of how to contain the spread across a random network is also well understood in principle. Each particular circumstance will have its own peculiarities, of course, and few real-world networks, certainly in the social and economic worlds, are completely random. However, in randomly connected systems, there is a critical mass, a critical proportion of the population as a whole, which has to be infected for the virus, or shock, to spread more generally across the population. Equally, if a critical proportion of the population can be immunized against it, the shock can be contained and will not percolate across the system as a whole. These principles underlie policies on, say, immunizing children against measles.

Systems in which the connections are not random but follow a power law have completely different properties. Here again, physicists have made an important contribution in recent years towards discovering and exploring such properties. So, too, have Steve Strogatz, an applied mathematics professor at Cornell, and Duncan Watts in the sociology department at Columbia, who famously discovered 'small-world' networks.

In networks in which the number of connections for the individual components follows a power-law distribution, the critical proportion of the population that needs to be infected for the entire system to become infected does not exist. In other words, in principle, if just one single component becomes a carrier of a virus or a shock or an idea, there is a risk that this will percolate across the entire population. Of course, the probability that this will happen is very small indeed, but it is greater than zero. If suddenly a substantial number of people become infected, the chances are higher that it will percolate across the system, but the possibility exists that the shock will be contained and fade away. In a purely random network of connections, neither of these properties are true. Either enough agents are infected for their number to be above the critical value for the shock to spread, or they are not, in which case the shock will simply die out.

Policies which try to contain percolation across such networks are very hard to implement. Unless each one of the very small number of highly connected individuals is targeted effectively and immunized, the virus will persist. Immunization at random of even a large proportion of the population will simply not succeed,

unless purely by chance not just some but all of the highly connected are caught.

These discoveries are all very recent, and much remains to be
done. Watts, for example, has investigated the implications for
the percolation of shocks across systems under different degrees
of connectivity of the network.* In other words, whether the
individual components are connected only loosely or whether the
network of connections is more dense.

The novel and powerful insight of all this work, however, is that
it is the structure of the connections between the component parts
which gives systems their distinctive and characteristic features.
Different types of networks, different structures, can give rise to
completely different characteristics for systems as a whole and, in
particular, to their vulnerability to shocks, to the possibility of
even the smallest shock being transmitted across the network as a
whole. In order to understand such systems, we do not necessarily
need to understand in great detail how each of the component
parts works, what rules of behaviour it follows – this is what, for
example, the economic theory of the firm tries to do – but we do
need to know how the component parts are connected. This is the
critical piece of information we need to understand the system.

The most dramatic recent practical illustration of the existence
of power-law behaviour, and in particular the failure to appreciate
that it exists, relates to financial markets. As we shall soon see, the
collapse of Long-Term Capital Management (LTCM) in 1998 was
brought about because of this. And the collapse almost, but not
quite, triggered the collapse of the world's entire financial system.

By way of introduction to the LTCM saga, consider the events of
September 1987, when the Dow Jones fell by almost a third in less
than a week, with just a single day showing a collapse of over 20 per
cent. Even now, the best part of twenty years after the event, there
are many competing *ex post* explanations, and no one is sure about

* D. J. Watts, 'A Simple Model of Global Cascades on Random Networks', *Proceedings
of the National Academy of Science*, Vol. 99, pp. 5766–71, 2002. An application of this
theory is my own paper 'Information Cascades and the Distribution of Economic
Recessions in Capitalist Countries', *Physica A*, vol. 341, pp. 556–68, 2004.

the cause. The only thing which is certain is that no external event took place which was on anything remotely near the same scale. There is no fingerprint, no smoking gun which points to the cause of this collapse. World war was not declared, the oil wells in the Middle East were not blown up. And yet, in the space of a single day, the largest companies in the world were suddenly declared to be worth 20 per cent less than they were the day before.

The stock-market crash of September 1987 is a dramatic illustration of how small events can have enormous consequences. The immediate causes of the collapse were small, so small, in fact, that even with the benefit of years of hindsight they have not been identified with any degree of confidence. Big changes may be very infrequent, but this episode tells us that they can be caused by small events.

At the time, the discovery that changes in the prices of financial assets – equities, bonds, currencies – follow power-law behaviour was a decade or so into the future, but we now know that a relationship exists between the size and frequency of price changes which is qualitatively similar to that between the size and frequency of biological extinction events.

Highly paid analysts in financial markets are always ready to articulate confidently and fluently why a major change has taken place, whether in the price of the stock of a particular company, the rate of exchange between two currencies, or whatever. But their record of trying to explain why something has happened after the event is very much more impressive than their ability to anticipate it.

In many ways, these accounts are no more than bedtime stories for grown-ups, serving both to console those who have lost money and to impart a delicious frisson of *Schadenfreude* amongst those who have not, for the stories are usually no more than fantasy. When it emerges that a company such as Enron or Parmalat – the latter wiping the smiles from the faces of many Europeans, smug at what they saw as the unique corruption of American capitalism – has falsified its accounts by incomprehensible amounts of money, it is obvious why the share price then falls.

However, much more often than not, the reasons remain very hard to understand. For example, interest rates are widely believed to affect exchange rates, but if, say, European interest rates are increased compared to those in the US, should the euro rise or fall against the dollar? The answer is that it could do either. One story is simply that investors can get a higher return on their money in euros, and so switch out of the dollar and into euros. But against that, an increase in interest rates can be interpreted as signifying a concern about the underlying strength of the European economy, giving investors reasons to think about selling euros and buying dollars. This is not a matter of pure theory, with the events of the past few years showing that both scenarios have actually happened. European interest rates really have been higher than American ones for a few years and, for a time, the dollar was strong against the euro and then became weak. Each time, interest rates were cited as one of the reasons for first the strength and then the weakness.

A most powerful illustration of what can happen when conventional thinking is confronted by reality, when truly dramatic events can occur for no obvious reason, is provided by the collapse of Long Term Capital Management (LTCM) in 1998. A full and graphic account is given by Nicholas Dunbar in his excellent book *Inventing Money*.

In September 1998, the world's largest single financial rescue package was announced. Not to save Russia, which had just gone bankrupt; not to save a massive industrial company; but to save an obscure hedge fund by the name of Long-Term Capital Management. Without the bail-out, the world's financial markets were on the brink of collapse.

Above all, the history of LTCM is the history of an idea, one which won the Nobel prize for its originators but which was proved wrong in a truly devastating way. In the late 1960s and early 1970s, a trio of American academics, leading successful but blamelessly obscure careers, discovered ways of applying concepts from statistical physics to financial markets. Fisher Black, described by one of his close friends as 'the strangest man I ever met', soon left academia to make millions at Goldman Sachs before his tragically early death. Robert C. Merton, son of the eminent sociologist Robert K. Merton, and Myron Scholes received the Nobel prize in economics in 1997 for their findings.

The work of Black, Merton and Scholes enabled the creation of today's trillion-dollar industry of financial derivatives. The basic idea of derivatives – so called because their value is derived from, or related to, that of an underlying asset – is very simple. As Dunbar points out, the concept can be found in the clay tablets of the Babylonians. Suppose an investor holds some IBM shares. He or she may worry that the price will fall. Someone else may think it will rise. A contract can be struck between them to trade the shares at a specified price at a date in the future.

The mathematics of pricing such contracts rapidly become very complicated, and the Nobel prize-winners appeared to have found a formula which could be applied very generally. Their work was distinctly original, but in the context of a very conventional way of thought. Like many innovative papers, their initial work was rejected by a number of journals before it found a home in the *Journal of Political Economy*. Based in Chicago, this distinguished and scholarly journal was strongly oriented towards free markets and the concept of equilibrium in economics, an idea at the heart of the derivatives model. For Black and his colleagues appeared to have found the missing link in the story of economies as vast machines, always in equilibrium, which had until then eluded the best minds in economics.

Their pricing model helped realize a long-standing dream of economists to construct a complete theory in which everything can be assigned a value in a market. Until this breakthrough on setting the prices of derivatives, conventional economists struggled with how to incorporate the future into their model. Now, it seemed that every possible state of the world, past, present and future, had a financial payoff associated with it. Every single market in the world, both present and future, could be brought into balance by the workings of the price mechanism.

Merton and Scholes were a driving force in LTCM. Indeed, in his autobiography provided to the Nobel Foundation, Merton depicted LTCM as the climax of his career. But the real culmination of these ideas was their empirical falsification and a $4.6-billion loss by LTCM. As Dunbar notes, 'The last seven days of LTCM's independent existence have a strange feel of their own. Thirty years of finance theory has proven itself useless. Billion-dollar track records and Nobel Prizes are now meaningless.' A major rescue operation had to be mounted by the Federal Reserve to prevent devastating consequences for the rest of the world's financial system.

The fund failed essentially because it embodied a view of the world in which big changes, big events, have big causes, and since we see very few large identifiable shocks to economies, certainly in the western world, it is then easy to believe that major changes in financial markets will happen with only exceptional rarity.

Here is what real statistical physicists Gene Stanley and some of his colleagues wrote about stock-market fluctuations in 2000: 'The probability distribution of stock-price fluctuations: stock-price fluctuations occur in all magnitudes, from tiny fluctuations to drastic events such as market crashes. The distribution of price fluctuations decays with a power-law tail and describes fluctuations that differ in size by as much as eight orders of magnitude.'

A certain amount of translation of this quote may be in order. First, stock-market changes, like the sizes of biological extinction events, take place across a very wide range of values, from small to extremely large. The phrase 'eight orders of magnitude' is used here in its precise, technical sense. To say something is an order of magnitude bigger than something else is a rather inexact phrase in everyday English. It simply means 'a lot'. In its scientific use, it means exactly ten times bigger. So eight orders of magnitude mean ten to the power of eight – ten multiplied by itself eight times, or one hundred million.

Second, there is a power-law distribution connecting the size and frequency of these events. Small changes in price are enormously more frequent than large ones but, every so often, exceptionally large ones can and do take place. It is this which the Nobel prize-winners, fixated by a world of order and equilibrium, failed to take into account in their complicated models,* and it is this which brought about the destruction of LTCM and the near collapse of the world's financial system.

The performance of the stock market is intimately connected to the performance of the companies which are quoted on it. A share which is owned in a publicly listed company essentially only has any value because of the present and future flow of dividend payments to which the owner of the share is entitled. Of course, there are many proximate reasons for buying shares in a company other than the dividends which flow from them. In particular, an investor may buy shares in the anticipation that they will rise in value but, without payment of dividends at some point, shares simply do not have any value. And to be able to pay dividends, the company needs to make a profit.

At this point, we might recall the stock-market booms of the late 1990s. Companies were floated on the market, and the value of their shares soared many times on the very day on which they first became available. Many of these dotcom companies never paid a single penny in dividends all their lives, but this is precisely the point. Without dividends, or at least the serious likelihood of dividends in the near future, these companies ultimately did not have a life. They failed. They became extinct.

Profits, and the dividends which flow from them, are the key to the value of company shares. An immediate health warning is essential here, for too much excitement can be bad for the nerves: readers who think that the secret of successful investing is about to be revealed will be disappointed. Shares in a company derive value from the profits of the company but, at any point in time, we do not know what the relationship between the two is. Two

---

* More technically, the LTCM models operated with the assumption, used in almost all the financial economics literature, that price changes follow a Gaussian and not a power-law distribution. Extremely large changes can occur in principle under a Gaussian distribution, but enormously less frequently than under a power law. For example, if price changes followed the Gaussian distribution, changes of the sort seen in September 1987 could only be expected to be observed at intervals of many millions of years.

companies may make their profit announcements on the same day. One may announce large profits, only for its share price to fall, because the majority of investors were anticipating even larger profits. The other may reveal much smaller profits, only for the price of its shares to rise, because many people were predicting an even worse performance.

The fact that different investors hold different views about what might happen may seem such a truism as to be not worth stating, but it is the reason why markets exist at all. If all investors held absolutely identical views and opinions, including, I may say for the benefit of any mainstream economist who happens to be passing by, their tastes and preferences towards risk, a market could not exist, for who would buy when the shares in a company were expected to fall in value? And who would sell when a capital gain were anticipated?

Bizarrely, a great deal of conventional economic theory assumes away the very differences of opinion which are such an integral feature of the real world. It is based upon the so-called 'representative agent', a single actor who calculates coolly and rationally the decision which will maximize his or her self-interest from now until the end of time. To the non-economist, this may seem incredible, but the words in the previous sentence are an accurate translation into English of the maths which underlies the theory of the representative agent in orthodox economics.

The representative-agent model obviously makes dramatic simplifications about reality. No one, not even the most devout believer in the economics of this approach, actually thinks that all economic actors really are identical or that they really do calculate the consequences of a decision from now until eternity, but all theories, of necessity, must simplify on a stupendous scale. As we noted in Chapter 4, a scientific theory abstracts from detail and tries to simplify the problem in order to be able to understand it better.

The eponymous hero in Kingsley Amis's novel *Lucky Jim* eventually comes to the realization that 'there were so many ways in which nice things were better than bad ones'. He plans to begin his celebration of this discovery with an octuple whisky. We will eschew such pleasure, for a clear head is needed. There are so many ways in which good theories are better than bad ones. One of the main ways is that the assumptions in a good theory, the simplifications it makes in order to comprehend reality, are reasonably realistic; another is that the theory should not leave out or be unable to explain key features of the issue being analyzed.

In the context of firm extinctions, standard economic theory does neither of these things. In reality, firms differ: they differ in their size, in their cost structure, in their capacity to innovate, in the ability of their management, to name only a few. And in reality, firms die. They die in great numbers. And we saw earlier from the Hannah and Fligstein research in Chapter 1 that even giant companies shrink and disappear.

More than 10 per cent of all economically active firms in the US become extinct each year. It is a distinctive feature of firms, and any theory of the firm should attempt to explain it. Conventional economic theory can only do so by positing an endless supply of completely unexpected shocks, for otherwise the perfectly informed, rational decision-making firm could never die. It would live for ever.

The American Office for Advocacy database contains information on just under six million companies in the US. These range in size from one-person businesses with an annual turnover of under $25,000 to Microsoft and General Motors. On average, over 600,000 of these companies disappear each year. Extinction is a striking and pervasive feature of firms.

In the same way that a pattern can be discerned amongst the apparently random fluctuations of biological extinctions, so, too, can we discover a structure to firm extinctions. Figure 10.1 plots over time the number of disappearances each year between 1912 and 1995 of Hannah's largest 100 industrial companies in the world in 1912, which we came across in Chapter 1. It is the raw data, in the same way that the number of extinctions per million years charted in Figure 9.1 in the previous chapter is the raw data on the history of biological extinction rates.

In most years, none of these corporate giants vanished as independent entities, so the most frequently observed value in this

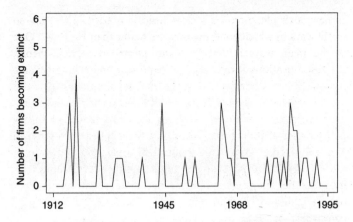

FIGURE 10.1 Number of extinctions of the world's largest 100 companies in 1912, annual basis, 1912–95.

chart is the value zero. In the remaining years, we see the irregularly spaced peaks which are qualitatively similar to Figure 9.1, which plots the size of biological extinction over time.

Both Figures 9.1 and 10.1 exhibit the very occasional large spike, indicating a high level of extinction, and quite a lot of smaller spikes, against the background of a typically low level of extinction. In the case of the giant firms, there are many years when none became extinct, so the plot of the data over time is flat at zero. This has never been the case with biological extinctions, which may make Figures 9.1 and 10.1 seem different. But again, to emphasize, they are qualitatively very similar. Both show typically low extinction levels, punctuated at irregular intervals by much larger ones.

Hannah's data spans just eighty-three years, so we have eighty-three data points along the bottom axis of the chart in Figure 10.1. The biological data in Figure 9.1 relates to seventy-seven separate geological periods, so we have seventy-seven separate points along the bottom axis in that chart. The Office for Advocacy data contains many more data points. Firms are grouped into different categories, by state, by industry (the economy is divided in this database into

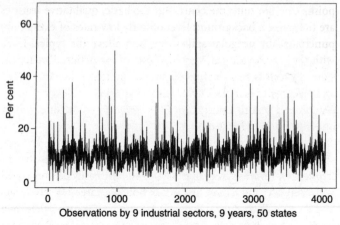

FIGURE 10.2 Percentage of US firms becoming extinct each year by industry and state.

nine sectors) and by year. The percentage becoming extinct in each of the separate groupings can be calculated. Figure 10.2 plots the information over nine years, so there are nine years, nine industrial sectors and fifty states, giving no fewer than 4,000 observations.

The average rate of extinction is just over 10 per cent a year, but there are occasions when virtually no firm becomes extinct and examples of when more than 25 per cent of all companies in a particular category become extinct in a single year, with occasional peaks of more than 60 per cent extinction. In the construction sector of the Californian economy, for example, during 1992 no fewer than 11,699 of the total of 56,351 companies ceased to exist, nearly 21 per cent of the total. In Kentucky in 1990, in the mining sector, 164 of the 798 companies failed in 1995, again over 20 per cent of the population in this sector. In the District of Columbia in 1989, 476 out of the 2,391 firms in the financial, insurance and real-estate sector failed, yet again 20 per cent. Extinction appears to be no respecter of place, date or type of economic activity.

The data plotted over time may at first sight seem different from Figure 10.1 because it is so much denser. There are so many more points to cram onto the chart. But the basic, qualitative features are the same: a background level of fairly low rates of extinction, punctuated by irregular spikes over and above the typical level, with the very occasional, very high rate of extinction. Exactly the same pattern is seen in firm-exit data in eight other developed western countries in the OECD which I analyzed with Corrado De Guilmi and Mauro Gallegati of the University of Ancona in Italy and published in *Physica A*.

We can plot the data in these charts in a different way which helps to bring out the relationship between the size of extinctions and the frequency with which they appear. This is more easily demonstrated with Hannah's 100 largest firms data, both the large US set of data and the OECD data for the eight developed economies other than America, requiring some rather technical prior analysis which need not concern us here.*

A standard way of presenting data organized by the frequency with which different values are observed is in a type of chart known as a histogram. We can see in how many years, in the Hannah data set, no large firm became extinct and record this frequency as a point in the data in the histogram. We go on to count in how many years one of the firms became extinct, two of them, and so on, up to the single year, 1968, in which the greatest number of firms, six to be precise, became extinct.†

The biggest single bar in the chart corresponds to years when none of the large firms became extinct. Reading across from the top of this to the left-hand axis tells us that this happened in just over fifty of the years between 1912 and 1995. To be precise, in fifty-four of these years, no firm which was one of the world's 100 largest in 1912 became extinct. Similarly, we can read across from the other

---

* This involves, amongst others, the Central Limit Theorem and the relative speeds of convergence of different statistical distributions to the general class of Levy distributions, of which the Gaussian is an important special case. Full details are in the relevant papers cited in the Further Reading section at the end of the book.
† Statisticians may immediately qualify any results obtained with such data on the grounds that there are only seven observations. But the fact that the results are very similar when the much larger US and OECD databases are analyzed suggests very strongly that the 100 largest firm results are valid.

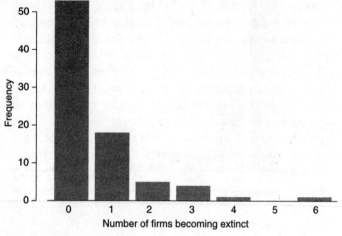

FIGURE 10.3 Frequency of annual extinction rates from 1912–95 of the world's largest 100 companies in 1912.

bars to see that in nineteen of these years one of them disappeared, in four years two did, and so on.

The firm-extinction data in Figures 10.1 and 10.2 already look very much like the extinction data of biological species plotted in Figure 9.1 in the previous chapter. Remember that this showed the percentage of total species becoming extinct per million years in each of the geological eras since around 550 million years ago (before present or BP). We went on to plot this biological data in a way that revealed the existence of the power-law relationship between the frequency and size of extinction events. The frequency with which an extinction of any size is observed falls away in proportion to the square of the size. If we compare two periods in which the rate of extinction per million years was twice as high in one than in the other, the smaller size will be observed four times more frequently than the larger.

In the previous chapter, we went immediately from the plot of extinctions over time to the plot of the power law and omitted the intermediate expository step of the histogram of the data. Figure 10.4

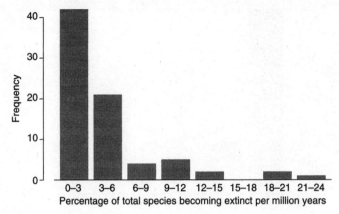

FIGURE 10.4 Frequency of extinction rates of biological species in geological eras since 550 million years BP.

fills in this gap and plots the Sepkoski frequency data of biological species extinction in a way which makes it as straightforward as possible to compare it with the firm data in Figure 10.3.

Figures 10.3 and 10.4 are not absolutely identical, but they present an immediate impression of similarity. The lowest extinction rate in the firm data was observed fifty-four times, and in the species data forty-two times. For the next lowest, the figures are, respectively, nineteen and twenty-one and, in each case, the higher is the number, or percentage, of extinctions, the less frequently it has been observed.

Not surprisingly, we can go on to reveal very similar power-law behaviour in the extinction pattern of firms as is found in biological species. Figure 10.5 below plots the Hannah data in the form of the log of the frequency with which an extinction size was observed and the log of the size of the extinction. This is exactly the form in which we plotted the species extinction data in Figure 9.2.* We see

---

* To be precise, Figure 10.5 plots the log of the rank of the size, so that the largest number is given rank 1, the second largest rank 2, and so on. This is a standard technique when the data series contains values of zero, which negates the possibility of using the logs of the raw data itself.

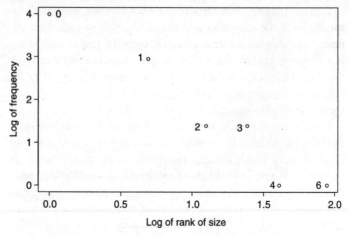

FIGURE 10.5 The frequency and size of the extinctions from 1912–95 of the world's largest 100 companies in 1912.

at the top left-hand corner of the chart the point labelled 'o'. This refers to the years between 1912 and 1995 in which none of the world's 100 largest companies in 1912 became extinct. Reading across to the left-hand axis tells us how many times this happened, in units of natural logs.

The striking feature of the chart is that the frequency with which different rates of extinction have occurred falls away as the size of the rate increases. For the point marked 'o', we read across to a high frequency on the left-hand axis and a low size on the bottom axis, in contrast to point '6' at the bottom, referring to the year when no fewer than six of these 1912 giants disappeared. In other words, as with the biological data in the fossil record, the number of times an extinction of any given size is observed is reduced as the size increases.

The relationship that describes the connection between the frequency and size of extinctions of biological species is the same as that which describes extinctions of companies. The timescales differ dramatically, but frequency and size are connected identically in both.

This startling result confronts us with a paradox. No biological species, with the exception of humanity, is able to anticipate the future and to plan its strategy accordingly. In reality, extinction is a pervasive feature for biological species, as it is for firms. Yet the people in companies *are* able to think about strategy, they are able to make decisions which will affect the ability of the firm to survive, and still extinction is a pervasive feature.

Throughout the book, we have encountered many examples of situations in which it is extremely difficult to penetrate the curtain of uncertainty which shrouds the future. From simple games to more complicated ones like chess to the real-life world of decision-making in business and politics, no matter how carefully researched and planned, the future consequences of decisions made today are frequently surprising. To have the intention of securing a particular outcome is usually no guarantee at all that it will be achieved. Intent is not the same as outcome.

In this chapter, we have seen a deep and subtle manifestation of this same phenomenon. Creatures cannot plan their evolution. The outcome of this process over biological time across species as a whole gives rise to a distinct pattern in the connection between the frequency and size of extinction events. Firms *can* plan their strategies, the way they intend to change and evolve, but we nevertheless observe the very same distinctive signature in their extinctions over time.

The implication is that it is as if – at last a useful and meaningful way in which to use the economists' favourite phrase! – firms acted at random, as if they were unable to act with intent and try to plan their futures. The massive uncertainty which often exists even in apparently simple situations means quite simply that intent is not the same as outcome. Firms try all the time to achieve favourable outcomes, but often they fail. And often they become extinct.

How can we examine this seemingly perverse conclusion even more thoroughly? For it does at first sight appear somewhat absurd, despite the many practical examples given in earlier chapters, to contend that giant firms are no more capable of planning and securing favourable outcomes with their changes of strategy than sunflowers are capable of deciding to grow feet better to follow the sun. In the next chapter, we start to approach this question in a scientific way. We do this by making assumptions, making simplifications of reality and constructing theoretical models which are capable of accounting for the key features of the empirical evidence on extinctions. These models can tell us even more about the true nature of the connection between intent and outcome, about the ability of firms to control their future destinies. We will eventually see in Chapters 12 and 13 that it is the structure of the connections between firms, the network across which the impact of firms' strategies percolates, which is the feature ultimately responsible for the patterns of extinction which we observe. First, however, we shall step back into the world of biology and consider the theories which have been developed to account for the new discoveries on the patterns of extinction.

11 Take Your Pick?

The crucial empirical discoveries of recent years regarding extinction in the biological world are two-fold. First, as might be expected given the random nature of evolution at the level of the individual species, the pace of extinction appears to vary over time in irregular waves. But, second, there is a clear pattern to this apparent disorder. A well-defined relationship exists between the size of an extinction event, what percentage of species becomes extinct in any given time interval and how frequently we observe it.

During the last few years, two ingenious and effective theoretical approaches have been articulated which account for these key features of the empirical evidence. Both can generate results which are consistent with the patterns which are observed, or rather deduced, from the fossil record. In many ways, the models' scientific power, whilst admirable in itself, is somewhat unfortunate, for the two theories imply dramatically different reasons for why extinctions take place at all.

The contrast mirrors an important theoretical debate within economics. Essentially, one of the approaches postulates that extinctions are caused solely by shocks that are external to the system. The other hypothesizes that extinctions arise quite naturally from within the system itself, from the network of connections between the individual agents and how this evolves over time. Of course, hybrid forms combining features of each can be constructed. In other words, we can have a theory in which extinctions arise from intrinsic features of the biological system, but which are augmented by external shocks. It is useful to describe the two approaches in their pure forms, however.

Economists make the same distinction in a number of important areas of economic theory, but they cannot resist dressing up the two approaches with rather fancy names. A phenomenon which only arises because of factors external to the system that contains it is described as being determined 'exogenously', and one which is determined purely within the system itself is 'endogenous'. As it happens, unlike several phrases in economics such as 'rational' and 'perfect', these have no ideological undertones and are quite useful.

A long-standing area of contention in economics, for example, is the so-called business cycle. This refers to the succession of expansions and contractions, of booms and busts, which we see in the economies of the developed world. Mainstream economics finds the existence of such movements a little disturbing. At the heart of such theory is a belief that the economy will naturally tend towards an equilibrium. This is a world in which either nothing changes at all or one in which any changes which might take place do so in a smooth and regular fashion.

This is most definitely not the case with the business cycle. The rate of growth of output, of GDP in economic jargon, is constantly changing from year to year. In the US, for example, from the late 1940s to the present day, the annual average growth rate has been around 3.5 per cent. But this conceals persistent fluctuations. The highest growth rate registered was nearly 9 per cent in 1950, and the largest fall was one of minus 2 per cent in 1982. This range of experience is typical of most western countries over the past half century.

Occasionally, however, we see much larger fluctuations. The collapse of output in North America in the Great Depression of the 1930s, when output fell by over 30 per cent in just three years, is perhaps the most famous example at one extreme. But the post-war world has not been immune to large collapses. Output in Finland, for example, fell by 15 per cent between 1990 and 1993. Post-war Japan, when output grew at an annual average rate of 8 per cent for thirty years and never fell below 3 per cent in any single year, is at the other extreme.

Moreover, year-by-year fluctuations in the growth rate of the economy show no signs of disappearing. Economic historians have strong evidence of their existence in the early years of the Industrial Revolution in Britain at the end of the eighteenth and start of the nineteenth centuries, and the recession in the US at the start of the twenty-first century brings home clearly the fact that these fluctuations have not disappeared.

Conventional economic theory gets round the conflict between theory, which postulates the existence of equilibrium, and the evidence, which shows persistent change, by assigning the existence of business cycles to external shocks. The precise nature of these shocks need not concern us here. Indeed, in practice, it has proved hard to identify what they might be. Nevertheless, the current mainstream theoretical approach to the business cycle in economics posits that it is determined exogenously, by shocks which are external to the system itself.

The idea that the business cycle is endogenous and is intrinsic to the workings of the western market-oriented economies is a distinct minority taste in economics. Exponents of this view include some of the most famous names in the whole of economics – John Maynard Keynes and Karl Marx to name but two – but the concept has been edited out of modern textbooks with a rigour that would meet the approval of Josef Stalin himself.

The within/without debate, whether key features of a system are determined inside or outside it, extends much further back in the corpus of western thought than economic theory. The great second-century theologian, Origen, developed a theory of redemption in which the deceased soul fluctuated between Heaven and Hell at irregular intervals. He was a stupendously productive writer, being the author of an alleged 6,000 books. His most famous, *On First Principles*, has perhaps the most marvellous intellectually snobbish title of any book ever written. Nevertheless, his theory of endogenous fluctuations in redemption was denounced as a serious heresy by the stern St Augustine and disappeared from orthodox teaching.

In the more mundane environment of biological extinctions, the American scholar Mark Newman has been at the forefront of developing theories of exogenously determined events. Currently

Professor of Physics and Complex Systems at the University of Michigan, he has made important, original contributions not just to extinction theory but to both theoretical and applied analysis of social networks, how people are connected to each other and how information, say, or disease spreads across a population, as well as to difficult problems in physics such as spin glass analysis.

Newman's basic model is beautifully simple, yet is able to generate the key observed features of extinctions in the fossil record. Everyone has heard of the theory that the dinosaurs became extinct because of some truly dramatic change in the external environment. A comet or asteroid struck the earth, or there may even have been a stupendous rise in volcanic activity from within the earth. In Newman's model, it is events such as these, over a very wide scale, which cause extinctions. The agents, the individual species, are not connected to each other. Extinction arises from the impact of external environmental changes on the ability of each species to survive.

Newman's model is populated by a number of different species. Each of the species is characterized by a stress tolerance level, which is unique to each species. In other words, any individual is able to put up with a certain amount of stress, or adversity, in its external environment. Once the overall stress level in which it lives exceeds this level, the species cannot cope and becomes extinct. We can think of the stress tolerance as indicating the fitness level for the survival of the species.

The transition is quite abrupt. At any level of external stress below its own particular tolerance level, the species thrives but, even when the stress increases only fractionally above the critical value, it disappears. Of course, reality would almost certainly be more nuanced than this, but the essence of any successful theory is the ability to make drastic but successful simplifications, and this is a simplification which succeeds.

The actors in Schelling's model, deciding what type of neighbourhood in which to live, are initially allocated at random. In the same way, in Newman's theory, each species is given a starting level of stress tolerance which is chosen at random. As in the Schelling game, the model then moves through time on a step-by-step basis. In each step, or period, the stress tolerance level of any given species is allowed to change. More precisely, the tolerance levels of a small fraction of the total number of species in the model is permitted to be altered in any single period.

We shall eventually see that this apparently innocuous refinement turns out to be very important in the context of social and economic systems but, at present, we merely note that it seems a reasonable assumption to make in this particular context. Evolution takes place at a fairly slow pace, certainly as far as the higher orders of species such as mammals are concerned, and this assumption is designed to reflect this fact. Only a small percentage of species evolve to a different level of stress tolerance in any given period, so any particular species will only evolve on average very infrequently. Suppose, for example, only 1 per cent of species evolve in any single period. This implies that, on average, any individual species will change on average every sixty-nine periods.*

Just three more rules are needed to make this model work. First, and most importantly, the level of external stress is chosen at random, and it is chosen afresh in every single period over which the model is solved. In other words, the external environment in which all species have to live is determined by a series of random shocks. Suppose, for example, that a huge meteorite struck the earth. In the Newman model, this would correspond to an extremely high level of external stress for all species. Not many species are equipped to survive searing heat, winds of thousands of miles an hour and tidal waves many miles high, to say nothing of the 'nuclear winter' which would follow such an event, as dust thrown up by the explosion

---

* At first sight, it might seem that, given there is a 1 per cent chance of changing in any given period, it will take fifty periods before the chance becomes 50 per cent. But the correct way to look at this is to realize that there is a 99 per cent chance of *not* changing. It takes sixty-nine periods and not fifty for this 99 per cent figure to be reduced to 50. In the first period, the chance of not changing is 99 per cent. In the second period, it is again 99 per cent, making the chance of not changing in the first two periods 99 per cent of 99 per cent, i.e. slightly more than 98 per cent.

remained in the atmosphere and screened out the sun for years afterwards.

The second rule decides how many species become extinct in each period. This is quite straightforward in the light of the above discussion. Every species whose individual stress tolerance level falls below the level of external stress becomes extinct immediately.

Finally, there is a rule for how extinct species are replaced. In the most basic version of the model, an extinct species is replaced immediately by another whose stress tolerance level is chosen completely at random. Of course, in this context, words and phrases such as 'immediately' or 'at once' should not be taken literally. Each single step, each period, in the model might correspond in real biological time to thousands or even hundreds of thousands of years, say. The word simply means that new species arise when old ones become extinct.

As with the Schelling model, these rules appear, and indeed are, quite simple: we have one rule which decides the fitness levels, or the stress tolerance, of the initial group of species; we have another which describes how these change over time; the actual level of stress faced by all species in their external environment is chosen by a further straightforward rule; and we prescribe both how species become extinct and how new species enter the model.

Despite the apparent simplicity, these rules give rise to complicated behaviour of the system as a whole. In the Schelling model, the level of preferences of the individual actors in the model is only mild, yet, overall, very marked segregation arises rapidly. This property of the system as a whole emerges from the interactions of the individual actors, or agents, in the model. It would take a von Neumann to be able to deduce merely from the individual rules alone that this would happen.

The Newman model, even though the agents are not connected to each other, also gives rise to unexpected behaviour of the system as a whole. The size of extinctions, the proportion of species becoming extinct in each time period, takes place in irregular waves over time, and the model generates the very particular relationship between size and frequency of extinction events which is observed in the actual fossil record.*

The model, like any other, has to be used with a certain amount of sense. Readers may recall, for example, something called Hooke's Law from high-school physics, which refers to the propensity of a spring to bounce back when stretched. Generations of school-children will have enjoyed stretching the spring in the classroom virtually into a straight line of wire, at which point, of course, it fails to spring back at all. In the same way, in the Newman model, if we were to draw the value of the external stress factor from a range which even at its minimum exceeded the highest level of stress tolerance of the agents in the model, no one could ever survive. However, this apart, the results seem very robust to the way in which the values of the stress factors are chosen.

The key features are also preserved when more sophisticated versions of the model are investigated. For example, a version has been developed which introduces different types of external stress. Each agent has a stress tolerance level for each of the types of stress and becomes extinct whenever one or more of the values of the external stress variables exceeds its relevant stress tolerance level. A different variant allows the fitness, or stress tolerance, to be inherited from that of a surviving species, which seems more realistic. Yet another permits a limited amount of interdependence between the species. In the basic versions, each species operates completely autonomously; its extinction has no consequences for any other. This rule can be relaxed, so that the disappearance of any particular species leads to the immediate extinction of its two nearest neighbours, regardless of their own levels of fitness.

The Newman model postulates that extinction is solely due to external events, that it is caused exogenously to the system.

---

* To be strictly accurate, the required relationship is observed over almost, but not quite, the whole of the range of extinction events. At very low levels of extinctions, when only a few per cent of species become extinct in a period, the size/frequency connection generated by the model deviates from the empirical evidence, but there are in any event practical difficulties of measurement when extinction rates in the fossil record are very low.

External shocks have certainly caused extinctions over the aeons of biological time – the evidence is too strong to ignore – but, as we have noted, not every extinction event, and not even all of the very largest ones, can be identified closely with a change in the overall environment.

The alternative view is that extinction arises solely within the system itself. Extinctions are posited to arise from the interactions across the network that connects the various species, as they co-operate and compete in complicated ways. In short, in this alternative view extinction is endogenous. As we have seen in previous chapters, in systems in which the individual components interact and influence each other's behaviour, the straightforward connection between causality and outcome is broken. Small-scale events, for example, usually only have small consequences, but occasionally they have much larger ones. On this view of the world, we do not necessarily need to identify the causes of major extinctions. They might arise, like the stock-market crash of September 1987, from trivial and unremarkable events.

A species can benefit from the existence of others. For example, many caterpillars are very particular about the food plants they will eat. Far from being omnivores they are what is known in demotic English as 'fussy eaters', so the more successful is the relevant food plant, the more the caterpillars survive and the more butterflies emerge of that particular species. Conversely, a decline of the appropriate source of food causes problems for the species. In contrast, species can compete with each other. We have seen examples where this is dramatic: one simply hunts and eats the other. But even non-violent species can find themselves competing for the same type of food, and the one less able to sustain the competition becomes extinct.

A model, yet again simple but devastatingly effective, of endogenous extinction has been developed by Ricard Solé and S. Manrubia. Like Mark Newman, Solé too spans a wide range of academic fields. With separate degrees in physics and biology, plus a doctorate in physics, he is currently research professor at the Catalan Institute for Research and Advanced Studies and a senior member of the

NASA-associate Center of Astrobiology. His recent work includes the evolution of gene networks, the origins and development of language, and work on astrobiology, with the ambitious aim of understanding the origin, evolution and distribution of life in the universe.

The way in which the model is put together might now be becoming familiar. A number of simple rules are specified. One is for how each solution of the model, or game, starts. In chess, this starting rule tells us how the pieces are lined up. In the Schelling model, it tells us that the agents are scattered at random across the board. The model then evolves over time, on a step-by-step – or period by period, as we can think of it – basis. So, in chess, at one step it is White to move, then Black, then White again. In the Schelling model, at each step an individual agent is called at random to decide whether or not to move. We also have rules which tell us what it is that changes in each period and how the changes take place. In chess, again, this describes what legitimate moves can be made by the various pieces. In the Schelling game, it describes both how an agent decides whether or not to move and where it moves to should it decide to move.

Solé and Manrubia begin by populating their model with a number of species or, in the model-building jargon, agents. The key feature of this approach is that the agents are connected across a network. Every single species in the model influences in some way the fitness for survival of every other species. Its behaviour either increases or decreases, by varying amounts, the ability to survive of everyone else.

The connections are described by what at its very simplest might be thought of as a spreadsheet, in Excel say. More mathematically inclined readers will think instead of a matrix. Imagine a spreadsheet with one hundred rows and columns, say. Each cell in this 100 by 100 spreadsheet contains a number, which summarizes the effect of the species in that row on the fitness of the species in that column. If a species benefits from the existence of another, this number is positive, and when a species is damaged by the actions of another, when its fitness for survival is reduced by the existence of the other, this number is negative. If there is no direct link between the two species, the number is zero.

So to start the game off, we specify the size of the spreadsheet and choose the starting numbers throughout the entire spreadsheet at random. The model then moves forward a step at a time in the way that is by now familiar. We specify a rule for what can change in each step and how it changes. In this model, or game, one of the cells in each row of the spreadsheet is chosen at random. This number is then changed to a new value, which is also chosen at random. Here we have the evolutionary process at work. In each period, each species undergoes an evolutionary change, which alters the impact it has on the ability to survive of another species.

We now have two rules. They tell us how at the start each species affects the fitness to survive of every other species, and how these effects change over time. A rule is needed to define when a species is deemed to have become extinct. This is very simple. We add up the impacts of all the other agents on the fitness for survival for any particular agent. If the sum of these impacts is greater than zero, it survives to live another period; but if the sum is less than zero, we specify that it has become extinct. We do this for each agent in turn, to see how many become extinct in any given period. In other words, we obtain information on the size of the extinction event.

Our final rule describes how extinct species are replaced. As in the Newman model, extinct species are replaced immediately. Niches which open up in the ecosystem are filled at once. A 'parent' species is chosen at random from those which survive. The newcomer is assigned the same values, themselves subject to very small random changes, across the cells of its row as those of its 'parent'.

This set of rules does the trick. The system as a whole generates extinctions whose size varies irregularly over time but which nevertheless has the distinctive pattern which we are looking for. A power law describes the relationship between the size and frequency of extinction events.

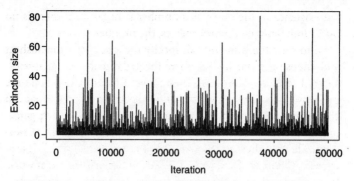

FIGURE 11.1 Percentage of species becoming extinct in each period in a typical solution of the model in which extinction is purely endogenous, i.e. determined within the system.

Figure 11.1 shows a typical solution of the model in terms of the percentage of species which become extinct in each period over time. An advantage of modern computer technology is that we can obtain such solutions very readily. In this case, we allow the model to run for as many as 50,000 separate steps, or iterations as they are known in the jargon of model-building.

Just as with the charts in Chapters 9 and 10 showing the actual biological fossil record and the extinctions of firms, we see irregular waves over time. For the most part, the percentage of species becoming extinct is low, but every so often we witness a large extinction and, very occasionally, we see an exceptionally large one. By casual inspection, for example, we can see that in six periods out of the total of 50,000, or approximately one hundredth of 1 per cent of the time, more than 60 per cent of all species became extinct.

In a way which may not be surprising by now, this data can be translated into the power-law relationship between frequency and size of extinctions which we again came across with both the real-life biological and firm data. Figure 11.2 plots the log of the frequency of the extinction events in the data in Figure 11.1 against the log of their size.

At the far left of the bottom axis, extinction size, the percentage of species becoming extinct, is small. Reading up from this, we see

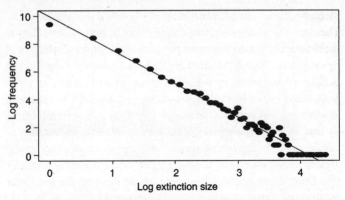

FIGURE 11.2 Frequency and size of extinction events in the model in which extinction is purely endogenous, i.e. determined within the system.

a dot which indicates how frequently the model generates an extinction of this particular size. Reading across to the left-hand axis from this dot, we see that the frequency is high. Conversely, the frequency with which very large extinctions take place, namely those at the far right of the bottom axis, is very low. There are a number of very large extinctions, but reading across to the left-hand axis of the chart, the frequency with which they happen can be seen to be very low.

So we have two abstract, theoretical models which are both capable of generating results that are compatible with the key empirical evidence on extinction in the biological fossil record. These properties emerge from the interplay of some deceptively simple rules about how the fitness for survival of species evolves over time and how extinct species are replaced by new ones.

The problem is that each offers a completely different explanation of why failure happens as it does. One is built on the assumption that extinction can only arise from external shocks. The other is premised on the view that failure is inherent in the ways in which species interact with each other. It arises from within the dynamics of the ecosystem itself.

Can these models translate into a social and economic context? And, if so, is there a way of distinguishing between them in terms of their effectiveness in understanding why failure is so commonplace? These are the issues which we consider in the next chapter. So far, the similarities between systems populated by biological species and by companies have dominated any differences between them. They both show the same deep patterns of extinction behaviour and they both imply severe limits to the abilities of their component parts, whether species or firms, to control their own fitness to survive by acts of deliberate strategy. However, there are, in fact, subtle but important differences which will enable us to place much greater weight on one of the theoretical models when we consider why firms, for example, become extinct and exhibit the particular patterns of extinction which we have described.

12 Resolving the Dilemma

The key facts on patterns of biological extinction can be accounted for by two completely different theoretical models. One assumes that extinctions are caused purely by shocks which are completely external to the system in which species live. The second postulates that extinction arises from the interactions of the species themselves. In other words, it is generated internally within the biological system itself.

In practice, a combination of the two is almost certainly responsible for the extinctions which we observe in the actual fossil record. As I write this chapter, considerable excitement has been generated by the claim that traces of a massive meteorite strike have been detected off the remote north-western coast of Australia. One of the very largest extinctions of species recorded occurred some 250 million years ago. The hypothesized date of the recently discovered impact crater is of a very similar antiquity, and so a possible external cause of this huge extinction has been identified. In general, however, extinctions will have taken place in practice through a mixture of external impacts and the competitive interactions between species.

In terms of the extinction of firms, as with biological species, it seems plausible that a model which incorporates both external and internal causes of extinction, exogenous and endogenous in the economic jargon, will be required. Firms are obviously endogenously connected to each other in terms of their commercial dealings. Equally, there clearly have been some important shocks leading to extinctions that have been completely exogenous, such as the Bolshevik seizure of power in the Russian revolution in 1917 and the expropriation of business assets or the nationalization of many mining companies in Europe in the late 1940s. What is not clear is the respective weight which needs to be given to the exogenous and endogenous mechanisms.

To recap, there are two key features of extinctions, whether of species or firms, which any theoretical account needs to be able to explain: first, the size of extinctions over time varies in irregular waves; and, second, there is nonetheless a distinct pattern which can be found in this seemingly random movement. There is a very particular relationship between the size of an extinction effect, or how many things become extinct within a given time period, and the frequency with which we observe it.

We can add a third fact as far as the disappearance of companies is concerned: the probability of failure, or extinction, is known to be highest when the company is first formed. It then falls away rapidly. After a short time, just two or three years, the probability of failure in any period of time is then unrelated to the age of the firm. This seems wholly plausible. Very basic mistakes are often made in the early months of the life of a company, such as incurring fixed costs involving premises and staff which are simply too high, or, a frequent cause of failure, ignoring the importance of cash flow. These weed out very quickly those least adapted to survive. Soon, however, the value of more experience falls to zero. No matter how long the firm has survived, beyond the initial danger period the probability of failure in the immediate year ahead is virtually the same.

Perhaps more surprisingly, there seems to be very little connection between the size of a firm, once the first few fraught years of existence are passed, and its probability of survival in any given period. Large firms are slightly more likely to survive than small ones, but the effect of size is distinctly limited. Leslie Hannah, the distinguished economic historian whose work we came across in Chapter 1, summarizes the evidence in rather dry, technical terms: 'To raise a joint-stock company's half-life by one year, it is necessary to increase its size by twenty-three times.'

The phrase 'half-life' may be recalled dimly from knowledge of radioactive elements, in which context it refers to how long it takes for half the particles of a radioactive element such as plutonium to decay and disappear. In the context of firms, it means: how long does it take for half of any given population of firms to become extinct? So we interpret Hannah's phrase as follows. We take a group of companies of one particular size or within a particular band of size. In order to raise the life expectancy over which half of them will disappear by just one year, we need a group which is on average twenty-three times bigger.

These facts support the view we have formed about the ability of firms to bring about a desired strategic outcome. After a short period of existence, once a set of very elementary errors have been avoided, there is almost nothing to be gained by further experience in terms of enhancing the prospects for survival. In addition, the benefits of being large extend only very weakly to the ability to avoid failure and extinction.

In a good detective story the clues to the denouement must be subtle rather than obvious. A hint that the external shock model might not work too well in a human context was given in the previous chapter. We mentioned there that the stress tolerances of the agents in the model only changed slowly. More precisely, the tolerance levels of only a small fraction of the total number of species in the model is permitted to be altered in any single period. We noted that this seemed very sensible when considering the evolution of biological species. The external environment is changing constantly, but in relative terms the pace of evolution is slow. This is most emphatically not the case with firms, or indeed any other human institutional structure. Even governments can respond with surprising alacrity, as the US administration showed after the 9/11 terrorist attack. Firms can alter their strategies very quickly.

This seemingly small change has dramatic consequences for the properties of the exogenous shock model of extinction. Once we allow the agents in the theoretical model to update their strategies with reasonable regularity, external shocks alone cannot generate the key features of firm extinctions which we observe.* In particular, the

* For interested readers, a number of technical papers giving details of these and other results are available on my website, http://www.paulormerod.com.

relationship between the size and frequency of extinction events no longer emerges from the interplay between changes in the external stress levels and the stress tolerance levels of the individual agents.

The results vary in subtle and unexpected ways, depending upon how we introduce the changes. The most obvious discord between theory and empirical evidence occurs when we retain the basic structure of the model exactly as described in the previous chapter but allow a bigger proportion of agents to update their strategies every period. For example, when only 1 per cent do so, on average, any individual species will change its strategy only every sixty-nine periods. So if we imagine, for example, that external shocks of any significance take place once every 1,000 years, say, then it takes on average 690,000 years for an individual species to undergo an evolutionary change.

However, it simply does not make sense to assume that only 1 per cent of companies change strategy every year. Firms constantly review their policies over the whole range of activities in which they interact with others, whether competitors, suppliers or customers. Even if we assume in the theoretical model that, say, only 20 per cent of companies change their strategies every period, the relationship between the size and frequency of extinctions breaks down. In reality, of course, even this pace of change is far too slow, and most firms make some changes to their policies in most years, but the model with this assumption can no longer give us the straight-line relationship between the logarithms of the size and frequency of extinction events.

We can try to rescue the external shock model in the context of social and economic systems by introducing variants of the model. One way of doing this is to let the agents be connected rather than operate purely in isolation from each other. We cannot, of course, make the network of connections too dense, for then we change the model from being one in which external shocks predominate to one in which the internal connections are mainly responsible for extinctions. But we can, for example, let the extinction of any given agent lead to those agents which are directly connected to it also becoming extinct with a given probability; or we can allow the extinction of any given agents to lead to those agents which are directly connected to it each having their fitness levels redrawn at random.

These and other variants make the model a little more realistic. In some of these, we once again obtain the straight-line connection between the log of the frequency of extinction events and the log of the size, but it has quite the wrong slope, being either markedly too shallow or too steep.

In contrast, the theory which attributes all extinctions to the competitive interactions of species translates remarkably well into the context of companies. Failures arise here from the complex ways in which the individual agents in the model interact. In an economic and social context, the key feature of this model is the spreadsheet, or matrix to give it its Sunday-best name, which informs us about how each individual agent's strategy impacts on everyone else. In its most basic form, every agent is connected to everyone else directly. In other words, the strategy of each agent affects the fitness for survival of every other one, though the impact is different for each one, and every other agent influences the fitness of any particular agent.

The developed economies of the west each have an extensive network of connections, especially the very large firms with each other. General Motors, for example, will almost certainly have dealings with software companies, computer manufacturers, banks and insurance companies, oil suppliers, producers of heavy industrial materials, and so on. Its direct competitors such as Nissan and Ford are obviously connected to the company. Their actions and strategies most certainly impact on GM's fitness for survival. So, as an approximation to reality at this level, the assumption made in the model in its most basic form of a completely connected matrix seems reasonable and not too dissimilar to reality. However, sticklers for realism in theoretical models can be reassured that this assumption can be relaxed fairly considerably and cells indicating no connection between any pair of agents can be introduced without affecting the overall properties of the model.

There are three distinct ways in which any pair of agents can interact with each other. Firstly, the activities of each could have a positive impact on the fitness for survival of the other. These agents, or species or firms, however we might think of them, exist in a co-operative relationship. As we noted in Chapter 8, contrary to the perception one would receive from examining an economics textbook, these positive relationships pervade the whole economy. Companies source their inputs from a wide variety of industries, and improvements in the quality of the goods and services bought in increase the fitness for survival of a company. Equally, if the firms to whom a company sells prosper, this, too, is likely to benefit the supplier.

We can think more generally in social contexts of relationships in which both agents can have a positive effect on each other. In the classroom, for example, a set of pupils who are well-behaved and eager to learn will boost the morale and motivation of most teachers. They in turn are likely to be keener and more helpful, and a system of positive feedback is set up in which everyone benefits. Social groups in which criminality is frowned upon and where respectability is the norm reinforce each other's behaviour through their everyday actions and everyone benefits from this self-policing of the community.

At the opposite end of the spectrum, we can think of relationships in which both agents have a negative impact on each other's fitness for survival. Staying with the example of crime for the moment, across the counties of the US crime rates per 100,000 population vary enormously. The rate of crime, whether all crime or subcategories such as violent crime, in the worst areas is not just higher but several *thousand* times higher than the most crime-free counties. Now, there are substantial differences in the socio-economic circumstances of an inner-city ghetto and the rural mid-west, say, but these differences are simply not on the same scale of magnitude as the differences in crime rates. It is hard to avoid the conclusion that the social norms of the areas magnify dramatically the differences that might arise purely through differences in socio-economic conditions. Economists Ed Glaeser of Harvard, Bruce Sacerdote and José Scheinkman found strong evidence of such social interactions in a careful study of crime across the US, published in the prestigious *Quarterly Journal of Economics* in 1996.

In a purely economic context, firms can interact in ways which damage each other's fitness for survival. They can, for example, become involved in a price war, the end result of which is simply to reduce each other's profits. A less obvious example was the decision by the food superstores in Britain to open on Sundays some fifteen years ago. Legislation had made it possible for the first time, but the companies hesitated. Nevertheless, as soon as one began Sunday opening, the others were compelled to follow for fear of losing market share. However, the total sales passing through these stores did not increase as a result, for shoppers already had ample opportunity to visit the stores during the rest of the week. The consumers were happier, but all the companies themselves ended up incurring higher costs, through the extra wages paid on the Sunday, for example, with no increase in sales.

Finally, we have the case in which pairs of agents interact in a positive – negative way. In other words, one benefits whilst the fitness of the other is reduced. This corresponds much more to the competitive view of the world of conventional economic theory. Firms within an industry are typically involved in a struggle for market share. By pricing, promotion, advertising, sourcing and innovation, each tries to steal a competitive march on the other.

In most mature consumer-goods markets, such as foods, beer, washing machines and cars, the overall size of the market changes only slowly. It is mainly determined by the overall economic environment, and there is little which an individual firm, no matter how large, can do to affect the total amount of spending in any such market. What happens in terms of pay rises, tax changes, unemployment and so on is much more important, but firms most certainly can influence the share of the market which they obtain. When one firm gains sales, this often involves a reduction in sales made by a competitor, so in these circumstances the two agents are connected in a positive – negative way.

The endogenous model is a very general statement of how agents interact with each other in a complex environment. The specific application is to biological species and firms, for in both of these cases we have clear empirical evidence of the key patterns in extinction against which the model can be judged. But we can think of it as applying in principle across a wide range of contacts. Agents, whether firms, individuals or governments, take decisions which have an effect on the fitness for survival of others in the system. Sometimes, agents are connected in a way which reinforces each other in a positive manner, and sometimes in a negative; and pairs of agents can be linked in a directly competitive fashion, when the gains of one are reflected in the losses of the other.

Once we allow the connections to evolve over time, to change not just quantitatively but also qualitatively, the key features of both biological and business extinction patterns emerge from the model. Failures occur in irregular waves over time, but from this irregularity we can extract a clear relationship between the size and frequency of extinction events. Large ones are less common than small ones. We can never know in advance what the size of the next extinction will be, but we can assign probabilities to the scale on which it will take place.

As we noted in previous chapters, conventional economic theory has a great deal to say about how it believes firms behave, but we have not needed to invoke any of this at all in order to be able to account for the key empirical features of firm extinctions. Not a single ounce of orthodox theory has been of any use. We do not need the hypothesis that firms act rationally in a way calculated to maximize their profits. We do not need the rule that price should be set equal to marginal cost, which we came across in the second chapter, and nor do we need any more sophisticated variant of this rule which theory suggests should be used in appropriate circumstances. The theories of perfect competition, imperfect competition, oligopoly, duopoly, monopoly, each purporting to describe company behaviour depending upon how many firms there are in a particular market, none of them are relevant. The whole panoply of the economic theory of the firm can be cast aside.

Instead, we simply describe how firms are connected to each other and how these connections evolve over time. The value that is taken by any one of these connections at any point in time tells us the impact of the strategy of a particular firm on another at that time. These values can be either positive or negative; a firm can either enhance or damage the fitness for survival of another firm. Economic theory, for all its apparent sophistication, only concentrates on the negative connections, attempting to inform us about how firms compete. It is, at best, only half the story. Understanding how the agents are connected in a system and how these connections evolve over time is far more important than worrying about the specific behavioural rules, such as profit maximization, which might be ascribed to the individual agents.

In the biological world, both the exogenous and endogenous extinction models in their pure form can account for the key patterns observed in the extinction of species, but the strictly endogenous model, in which firms are connected to each other and have impacts on each other's fitness, translates far better into socio-economic systems than does the strictly exogenous one.

However, as we have noted several times, in the human world of social and economic organization, in practice failure and extinction almost certainly arise from a combination of endogenous and exogenous factors, of external shocks and the purely internal interactions of the component parts of the system. The internal network of connections and how it evolves over time are the most important causes of extinctions, but external shocks will often play a role as well. We now explore the implications of making the model even more realistic by introducing external shocks into the self-generating explanation of extinction.

13   Why Things Fail

In practice, of course, almost any system will both exhibit evolving interactions between its component parts and experience shocks which are completely external to the system. The shocks could affect all agents equally or the impact could vary across agents. The most obvious example in a biological context is a meteorite strike on the earth or the explosion of a super-volcano which has a major impact on climate. Less dramatically, we could think of the appearance of a different species in a settled environment, either driven there by necessity or introduced in an artificial manner by humans. The potential list of examples in human structures and societies is too numerous even to begin to list.

Adding this feature to the model of endogenous extinction, or failure, does not alter its ability to account for the key observed features of extinctions. Of course, like any model, it can be tested to destruction by making the external shocks so large that they come to dominate completely the internal interactions between agents. That aside, however, the model can be made more realistic by adding random external shocks.

When we do this, we obtain an insight into why it is so difficult in a social and economic context to draw firm conclusions using conventional ways of thinking. An important standard way of analyzing evidence in the social sciences is precisely to try to relate changes in the feature of interest, in this case the size of the extinction, to one or more specific factors. So, for example, we may try to account for variations in crime rates or unemployment, say, across localities at a point in time by correlating them with a set of socio-economic indicators. Crime, for example, might be held to vary with the proportion of families in poverty, or the number of single-parent families, or the levels of policing, or the sentencing policy of the criminal justice system.

Such studies proliferate throughout the social sciences, but their quality varies dramatically. Some are little more than ideological statements masquerading as research; advocacy research, as it is usually called. Equally, many are serious studies by scholars motivated purely by a desire, in the best scientific tradition, to understand the world a little better.

Undoubtedly, our understanding of the world has been improved by such research, but it remains imperfect. We only perceive the world through a glass darkly by these means. Conventional thinking offers an account of such impacts which is at best incomplete and at worst positively misleading. The effect of any given changes in a set of variables can vary substantially, depending upon the particular circumstances in which they are introduced. Quite frequently, small changes in one will have only small consequences for the other, and large changes will have large effects, just as common sense suggests they should. Sometimes, however, a change of identical magnitude which in a different context only had a small impact will now have a large effect, and occasionally it will have a really dramatic one.

An example of the failure of conventional thinking is given by the so-called Phillips curve. Named after a British engineer who made the discovery in a carefully argued article in 1958, this curve describes the relationship between the rate of inflation in an economy, the speed at which prices are changing, and the level. There are various ways of measuring this latter in practice. In the original article, Phillips used the unemployment rate, on the eminently plausible grounds that in recessions, when spending is cut back, workers are laid off and unemployment rises, and in booms unemployment falls. So it is a good proxy for the buoyancy or otherwise of demand.

Standard economics, in the form in which it is taught to most students and in which it is used by policy-makers whose studies of economics ended many years ago, postulates a connection between the price of a product and the demand for it. The Phillips curve can be thought of as a generalization of this principle to the level of the economy as a whole. The higher the level of demand, the higher will be prices. The specific form which is postulated is to say that the rate of change of prices, in other words inflation, will be connected to the level of demand; and, in particular, inflation will be connected in a negative way to unemployment, the higher being the rate of unemployment, the less will be the pressure of demand.

In my book *The Death of Economics*, first published in Britain in early 1994, I pointed out that in developed economies the Phillips curve appeared to be of little use either in forecasting inflation or in trying to control inflation by measures which affect the level of demand, such as interest rates or tax changes. There did indeed appear to be periods when a Phillips-type relationship existed between inflation and unemployment, but the position of the whole curve would suddenly shift, with little warning and for no apparent reason. In the US, for example, a simple plot of inflation against unemployment showed a reasonable Phillips curve for the years 1955 to 1971, then a sudden shift in the position of the curve. Another curve described the years 1974 to 1984 well, only for a further shift to take place in 1985. For some reason, this particular example attracted the scorn and ridicule of many of the conventional economists who reviewed the book. One entirely typical reviewer dismissed my arguments as 'naïve'.

The book was written during 1992 and 1993. I argued that the idea that there is a trade-off between inflation and unemployment which can be used for policy purposes is wrong. Small, unidentifiable events make the position of the curve which describes this trade-off suddenly shift. In the decade or so since then, the relationship between inflation and unemployment that is supposed to exist has broken down in spectacular fashion. In 1992, inflation in America was 3.0 per cent and the unemployment rate was 7.5 per cent. In 2004, they were 2.3 and 5.5 per cent, respectively. In Britain, inflation was 4.2 per cent and unemployment 9.8 per cent in 1992. The 2004 figures were 1.5 and 4.7 per cent. In other words, unemployment in both these economies has fallen dramatically and inflation has fallen as well, completely contrary to the conventional way of thinking.

Of course, as we discussed in Chapter 10 in particular, the relationship between the size of an event and the magnitude of its consequences is not completely random. Large shocks, say, will usually have bigger impacts than smaller ones. However, in general, in the sorts of networks that connect individual agents in social and economic systems, we can never rule out the possibility that the smallest, most obscure shock will turn out to have massive implications. The probability is very low, but it is greater than zero. It can and does happen.

The endogenous model of extinction that accounts for the patterns of company failures we observe has this property. We can apply shocks, of varying sizes, at random intervals to this model and examine the consequences. Figure 13.1 overleaf gives an example of the relationship between the size of the external shock that is applied to the model and the size of the extinction which is observed. For interest, we also show the effect of a shock on extinctions in the period that follows it.

Clearly, there is a positive relationship between the size of the shock and the size of the extinction which occurs. Each cross in the chart indicates a particular set of values in the solution of the model which was observed. Again, reading across to the left-hand axis we see the level of the shock, and down to the bottom axis the extinction size associated with this shock. There is a whole mass of points down at the far left-hand side of the chart, where both shocks and extinctions were relatively small; and sloping upwards across the chart are observations where larger shocks were linked with larger extinctions.

The relationship, however, is far from perfect. Larger shocks do indeed play a role in causing or exacerbating extinctions, but the largest shocks are not necessarily coincident with the largest extinctions, many of which coincide with shocks of the smallest size. The scatter of points across the chart just above the zero shock size tells us this very clearly. Smaller shocks are more numerous and so statistically more likely to coincide with a large extinction resulting from purely endogenous causes. Moreover, larger shocks do not necessarily lead to the largest extinctions

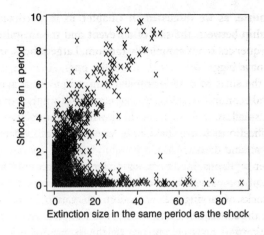

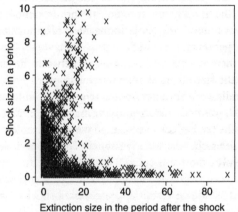

FIGURE 13.1 Plot showing (a) relationship between shock size and extinction size for a typical solution of the model; (b) for the same solution, plot showing shock size and extinction size in the period after the shock.

because they may strike at a time when the system is particularly robust.

In the living, constantly changing economic and social worlds, the connection between the sizes of events and the magnitudes of their effects is no longer routine and mechanical. Small changes often have small consequences, but occasionally these are large and, from time to time, dramatic. Equally, large changes sometimes have large effects, but they may also make surprisingly little difference to the eventual outcome.

The model of both biological and corporate extinction patterns illustrates these points clearly. From a scientific perspective, the model is very successful. It is based upon a small number of apparently simple rules, but these rules give rise to properties of the model which are hard, if not impossible, to deduce from inspection of the rules themselves. Crucially, these properties are consistent with key observed features of the phenomena we are trying to explain.

We can go on to use the model to revisit the important question of the gap between intent and outcome in the actions of individual agents. Again, although the specific illustration is in the context of the extinction model, the implications can readily be generalized to apply to any system in which the agents, individuals, firms or governments interact with each other across networks of connections and in which these connections change over time.

Specifically, we want to examine the extent to which agents are capable of systematic learning about the environment in which they operate. We have already seen in the very simple contexts of games such as chess and the Prisoner's Dilemma that a certain amount of learning appears to be feasible, but the bulk of the game's secrets remain beyond our knowledge. A very similar conclusion can be drawn from the game described by the successful theoretical model of extinction.

A great advantage of a theoretical model is that we can create artificial worlds. In other words, we can change the rules of behaviour and see what happens. In the basic form of the model discussed above, we have been assuming that the effect of strategies evolve purely at random. No one has any foreknowledge of either the impact of his or her strategy on the fitness of others or of the effect of the strategies of others on their own prospects for survival.

We can create variants of the model in which we deliberately assign varying amounts of knowledge to the agents. In other words, we can allow agents some element of control over the ways in which the impact of their strategies evolves. Instead of having these updated purely at random, as in the basic model, we can permit agents to exercise some level of successful intent and to alter the outcome of what would otherwise be a random change in their own favour. We can then observe what happens to the properties that emerge when the model is solved over time. In particular, we can discover what degrees of knowledge are compatible with the key patterns on extinction that we observe in reality.

Two features of behaviour in the model can be controlled in this way: first, the proportion of agents who acquire such knowledge; and, second, the amount that they obtain. The results of both are revealing.

One aspect of the results will gladden the hearts of economists of all persuasions. There are very considerable returns to acquiring knowledge, for even a small amount leads to a sharp increase in the average age of agents at extinction. Indeed, we find that as both the amount of knowledge available to agents increases and as the number of agents capable of acquiring such knowledge rises, their lifespan begins to approach the limiting, full-information paradigm of conventional economic theory in which agents live for ever. So if you can acquire knowledge about the impact of strategy, it is distinctly valuable.

To illustrate this, suppose that, without any knowledge or learning at all, agents live on average for 100 periods. This is very much an average, for as we have seen some will live only for a very short time, whilst a few will go on just like Methuselah, but the random changes of the impact of strategies in the complex web which connects agents lead to them living on average for 100 periods. Of course, we use 100 purely for descriptive purposes, and this should not be interpreted in any way as implying that in reality this is a typical lifespan. The figure is chosen as a benchmark for illustration so that we can easily describe the increase that takes place once different amounts of knowledge are made available to agents.

We adopt another descriptive convention regarding the maximum amount of knowledge that an agent can possess. Suppose an individual or firm knows absolutely everything there is to know about another agent. Suppose further that this individual has acquired complete knowledge about the impact of changes in his or her own strategy. In other words, we have a hypothetical omniscient being, which will never die, straight out of full-information, maximizing economic textbooks. For illustrative purposes, we can say that this individual will adjust so that the impact of the strategies of everyone else on it will all take the value of +1. Everyone will contribute to increasing the fitness of this particular agent. In contrast, if we assume the agent has no knowledge at all and so choose these contributions at random, they can be both positive and negative, and the impacts will on average be equal to zero. At one extreme, then, an all-knowing agent will contrive that all the cells which impact on its fitness for survival will take the value of +1 and, at the other, an agent with no knowledge at all will have cells which on average take the value of zero.

With these descriptive conventions in mind, suppose now we allow just one agent out of every 100 to acquire knowledge and adapt to the impact of the strategies of others on it. If it is able to acquire merely 5 per cent of the total possible – to increase the average value of each of its connections from zero to 0.05 – on average it survives for 140 periods. In other words, even this minimum amount of knowledge enables its lifespan to be increased on average by no less than 40 per cent. Of course, this will not always happen. There is no guarantee, of course, that such an agent will not be made extinct very quickly. In the same way, we have seen that sexually reproducing genes have an advantage over asexually reproducing genes, but these, too, may fail to survive. At a value of 0.10, just 10 per cent of the theoretical maximum of an all-knowing creature, the average age at extinction is 210 periods. The gains then rise even faster. At the values of 0.20 and 0.35, respectively, averages increase to no less than 540 and 2,050 periods.

Once we allow more than 1 per cent of agents to be able to acquire and act on knowledge, the gains are even more spectacular. Their directed behaviour reinforces each other and a positive feedback loop is created. So if half the agents have these powers, for example, with just 5 per cent of the theoretically possible maximum amount of knowledge, the average life is 160 periods. With 10 per cent it is 330 periods and with 35 per cent it rises to as much as 29,000. As more and more agents are allowed to acquire more and more knowledge, the average lifespan begins to approach infinity and agents live for ever.

We can understand now why marketing departments are so important in so many companies. They are ravenous for information about their customers, both actual and potential. Even a tiny bit of genuine knowledge goes a very long way. As Lord Leverhulme, who built the company from which the global consumer-goods company Unilever eventually emerged, allegedly said, 'I know that half of my advertising is wasted. I just don't know which half.' Just a small amount of insight into which half it is could save huge amounts of money over the years.

Agents with complete knowledge live for ever, so by definition we do not observe any extinctions in such circumstances. This is self-evidently unrealistic, because corporate extinctions take place all the time. In contrast, when the model operates as if agents have no knowledge and the values of their connections are updated at random, the model can replicate the key facts on extinction.

We noted in Chapter 8 the two extreme postulates that could be made about the cognitive abilities of individual agents to gather and process information and turn it into knowledge: the full-information rational maximizer of economic theory, and the agent acting as if it had zero cognitive ability. In terms of being able to replicate the observed features of firm extinctions, the latter is a much better description of reality than the former. Versions of the model in which agents are ascribed powers even remotely approaching those of the rational maximizer are completely unable to generate the key empirical features of extinction patterns. Scientifically, they fail.

Suppose, for example, we permit all agents to acquire just 1 per cent of the theoretical maximum amount of information. In other words, they are very similar indeed to the extreme of the zero cognition agent. In this case, the model is still successful; extinctions follow the required power law. But even at 5 per cent of this maximum level, the patterns of extinction which emerge are starting to deviate from the empirical evidence, and at 7 or 8 per cent, the dissonance is very clear.

If instead knowledge is concentrated on a subset of all agents, more can be imparted to the fortunate few. With just 1 per cent of the total population having the ability to influence the impact of strategy in a predictable way, their level of understanding can rise to some 50 per cent of the potential maximum. But if even just 10 per cent are in this position, they cannot be given more than some 20 per cent of the maximum before the model no longer replicates the key facts about extinction.

In short, despite the ability of humans and human institutions to act with intent, in reality it is as if they operate close to the paradigm of the agent with zero cognitive ability. They do not have to mimic it completely, and a small amount of ability to translate intent into desired outcome is compatible with the evidence we observe, but no more than that.

In reality, things do not live for ever. The largest firms eventually become extinct. In terms of political structures, Hitler boasted that he had created a Reich that would last for a thousand years. After twelve years, it had disappeared. Even in the mid-1980s, the Soviet Union appeared immovable. Very few people indeed could imagine a world order without it. Yet, by the end of that decade, after a life of a mere seventy years and despite its unparalleled apparatus of internal repression to secure the regime, despite its immense arsenal of nuclear weapons, it had gone. The Roman Empire lasted much longer but, eventually, it, too, fell.

Each of these monumental political events has many causes. Chou En-Lai, of the Chinese Communist Politburo, once famously remarked that it was too early to know the consequences of the French Revolution of 1789. Even some 2,000 years later, historians still cannot agree on the exact sequence of events that brought down the Romans, the most dominant empire the world had seen until the advent of the American empire in the late twentieth century.

This is exactly what is implied by the model of extinction. Even in a fairly simple model, it becomes hard and indeed virtually impossible to trace the precise reasons why an agent becomes extinct. The more knowledge we assign to agents, the easier it becomes both for them to survive and for us to understand their evolutionary patterns.

The clear implication of this abstract theoretical model is that agents, firms, individuals, governments have very limited capacities to acquire knowledge about the true impact either of their strategies on others or of others on them. The model passes stringent scientific tests of validation. The properties that emerge from it, which are not at all obvious from a description of its component rules, accord very well with the subtle but clearly defined patterns of extinction which we observe. And we have seen that even the world's largest firms are capable of making huge mistakes about the possible effects of their strategies.

To repeat a key phrase which needs to be hard-wired into the brain of every decision-maker, whether in the public or private sector, intent is not the same as outcome. Humans, whether acting as individuals or in a collective fashion in a firm or government, face massive inherent uncertainty about the effect of their actions. Whether it is the great characters of tragedy or giant corporations such as Microsoft, the future remains covered in a deep veil to all. Species, people, firms, governments are all complex entities that must survive in dynamic environments which evolve over time. Their ability to understand such environments is inherently limited.

These limits are a fundamental feature of the systems we have discussed, whether biological or whether in the realm of human social and economic organization, in which the individual agents are connected through networks which evolve over time. These limits can no more be overcome by smarter analysis than we are able to break binding physical constraints, such as our inability to travel faster than the speed of light. This is why things fail.

14 What Is to Be Done?

The implications of our analysis of failure and extinction appear to inhabit a totally different world from that of Dr Pangloss. A character in Voltaire's *Candide*, he has become famous for his boundlessly optimistic view of the world regardless of the objective circumstances. He teaches 'metaphysico-theologo-cosmolonigology', a brand of metaphysics promoting the philosophy of optimism. Pangloss states that the world is the best of all possible worlds, in which all is well and all is for the best. Though venereal disease almost kills him, though the Spanish Inquisition hangs him (unsuccessfully) and though he witnesses the misfortunes of his fellow men, Pangloss refuses to recant his optimism, a role model indeed for conventional economists, with their devotion to maximizing behaviour and equilibrium! Fortunately, it should be said, throughout the book the hero, Candide himself, compares Pangloss's theory with the empirical evidence and finds it sadly wanting.

Yet humanity is not completely powerless in the face of the Iron Law of Failure. There are positive attitudes, positive steps that policy-makers, in both the public and private domains, can take. Moreover, failure at the individual level can paradoxically be beneficial for the health of the system as a whole.

Like John von Neumann, whom we met in Chapter 5, Joseph Schumpeter and Friedrich Hayek were two massively original thinkers from central Europe who, in the 1930s, took permanent refuge in the west. Schumpeter became Professor of Economics at Harvard in 1932, where he remained until his retirement in 1949 and death the following year. Hayek took up a chair in economics at the London School of Economics in 1931, moving to Stanford in 1946 and then to Chicago in 1950, where he stayed for over a decade.

Both Schumpeter and Hayek were severely critical of conventional economic theory and of its ability to say very much of value about how the world actually works. Both carried out powerful and highly innovative work which has never been absorbed into the economic mainstream. Both, especially Hayek, have large and devoted intellectual followings, which are growing rather than shrinking over time. Both saw, decades ahead of their time, capitalist economies as the type of system that has been the focus of this book and which is now at the frontiers of analysis in a wide range of disciplines. Both believed that it was precisely the complex interactions between the component parts that led to the pervasive existence of failure. And both realized that failure at the level of the individual component, the individual firm, could be beneficial to the system as a whole.

Schumpeter was dismissive of the prevailing view in economics in his day that 'perfect' competition was the way to maximize economic well-being. We met this concept in Chapter 5, and it still remains the desirable market structure not just according to the economics which most students learn even today but in the economics used by many regulatory authorities around the world. Under perfect competition all firms in an industry produce exactly the same item, sell it for the same price and have access to the same technology.

Schumpeter saw this kind of competition as relatively unimportant, writing: '[What counts is] competition from the new commodity, the new technology, the new source of supply, the new type of organization … competition which … strikes not at the margins of the profits and the outputs of the existing firms but at their foundations and their very lives.' In other words, he believed that understanding the process of change, how firms evolve over time, was essential to understanding how modern economies prosper.

Schumpeter argued on this basis that some degree of monopoly was preferable to perfect competition. Competition from innovations, he argued, was an 'ever-present threat' that 'disciplines before it attacks'. He cited the Aluminium Company of America as an example of a monopoly that continuously innovated in order to retain its monopoly. By 1929, he noted, the price of its product, adjusted for inflation, had fallen to only 8.8 per cent of its level in 1890, and its output had risen from 30 metric tons to 103,400.

Hayek attacked conventional economic theory at an even deeper level. His achievement was to show that, *theoretically*, market-based economies are inherently superior to planned ones. Paradoxically, conventional free-market economic theory cannot demonstrate this result. Indeed, it is equally relevant to the idealized world of both the central planner and the pure free market.

Hayek demonstrated that desirable social and economic outcomes arise not merely from the actions of isolated individuals, which is the postulate of orthodox economics. Rather, they are the *joint* product of both individual actions and the institutional framework in which individuals operate. Individuals and institutions arrive at satisfactory outcomes by a process of evolution and competition. Rigid, centralized planning operating under a fixed institutional structure is the very antithesis of what is required.

Over two hundred years ago, Adam Smith gave examples of particular industries, such as bread baking, in which the self-interested actions of producers responding to incentives lead to benefits for everyone. Economic theory in the twentieth century was able to show that this could lead to a desirable outcome not just in a few but in *all* markets at the same time. Supply and demand would balance everywhere, and there would be no unused resources. An efficient, overall outcome would prevail. We have already met this model of 'general equilibrium'.

General-equilibrium theory is often seen as the crowning achievement of economics in demonstrating the superiority of free markets over planning. But, as Hayek realized, the theoretical implications of general equilibrium are that a centrally planned economy can be at least as efficient as a free-market one, and may even be superior. The ability of individuals and firms to both gather and process information in the model of general equilibrium is so strong that, in principle, a central planner in a socialist state might well be able to satisfy them more readily than a more decentralized, market-based decision-making framework.

During the 1940s and 1950s, the concept of planning was very fashionable in the west. It was also during these decades that the first really powerful mathematical results on general equilibrium emerged. A number of academic articles at the time demonstrated that an omniscient socialist planner, by using the price mechanism as a way of deciding how resources should be allocated, could achieve results identical to those of an equally idealized free-market economy, but with a more egalitarian distribution of income and wealth. In other words, in the world of general equilibrium, the great theoretical achievement of conventional economics in the twentieth century, socialism could be better than capitalism. Socialist planning in theory could be just as efficient as free enterprise, and at the same time more equitable.

It was Hayek's genius to offer a completely different, and much more realistic, view of how economies operate. In a Hayekian world, decentralized decision-making by individual agents is *unequivocally* superior to central planning. Indeed, a central plan may well be the worst possible institutional framework an economy could have. The decentralized, market-oriented model may not give the very best result, the optimal outcome, for in most circumstances we have no way of knowing what this is, but it delivers a satisfactory outcome, which benefits most or all of its component agents.

The visions of the world articulated by orthodox economics and by Hayek are fundamentally different. Conventional theory describes a highly structured *mechanical* system. Both the economy and society are in essence gigantic machines, whose behaviour can be controlled and predicted. Hayek's view is much more rooted in biology. Individual behaviour is not fixed, like a screw or cog in a machine is, but evolves in response to the behaviour of others. Control and prediction of the system as a whole is simply not possible.

Interestingly – and how unlike most modern-day economists! – Hayek understood and admired the achievements of other intellectual disciplines. Anthropology attracted his particular attention, and of all the social sciences he regarded this as the one which produced people who thought in a sensible way about the development of society. For Hayek, 'an economist who is only an economist cannot be a good economist'.

The complex interactions between individuals give rise to *inherent* limits to knowledge of how systems behave at the aggregate level. No matter how smart the planner, no matter how much information he or she gathers, there are inescapable limits to how much can be known about the system.

Vernon Smith gives a practical illustration of the limits to knowledge in his brilliant Nobel lecture, published in the June 2003 *American Economic Review*.* Airline route deregulation in the US has led to the emergence of the so-called hub and spoke system. There are few direct flights between cities, and most journeys involve a change at one of the small number of 'hub' airports. Smith describes this as an 'ecologically rational' response. Significantly, as he points out, *no one* predicted in advance that this institutional structure would evolve. This is not because airlines were stupid; it is because customers did not know themselves in advance that this was the system they preferred. They were not cogs in a machine following fixed rules of behaviour. Their preferences evolved towards a system with which they were satisfied through a process of market experimentation. Following deregulation, different types of route structure were tried, but the hub and spoke evolved as a reasonably efficient outcome which worked.

This particular institutional structure was discovered through a process of evolution and competition. Another example is provided by the evolution of a single currency in the US, and the marked contrast of this experience with the attempt to foist a common currency on Europe by means of central planning. Successful institutions, or rules of the game, are not set by central planning diktat. Instead, they evolve. The US dollar bestrides the world. Acceptable in every country, fervently desired in most, it is a potent symbol of American economic strength and power. But it was not always so. Indeed, it was less than a hundred years ago that the US Congress established

---

* V. L. Smith, 'Constructivist and Ecological Rationality in Economics', *American Economic Review*, 93, pp. 465–508 (2003).

the Federal Reserve Bank, the American equivalent of the Bank of England or European Central Bank.

For most of America's history, from the English colonies established on the seaboard fringe of the continent over four hundred years ago to the early twentieth century, a wide and at times bizarre range of different types of money circulated within the United States itself. The monopoly of the dollar there is a comparatively recent event. The American republic was established in 1783, but as late as the mid-nineteenth century, a tremendous array of different types of money circulated in America, with states and banks being free to issue their own notes. As late as 1860, there were some 9,000 different kinds of privately issued dollar bills circulating, around a third of which were counterfeit. On no less than six occasions in the first half of the nineteenth century, Congress passed acts allowing foreign coins – French, Spanish, British – to circulate as legal tender.

Two attempts to establish a central bank in the US, of the kind with which we are now all familiar, failed. Prosaically called the First and Second Banks of the United States, both had short lives which had ended by 1840. America then waited until 1907 before the Federal Reserve Bank, with us today, was established.

The US emerged as the most important economic power in the world during the course of the nineteenth century. By the First World War, the American economy was over twice as large as that of imperial Britain, the second economic power, and living standards were distinctly higher. Yet these stupendous developments took place against a background of a baffling plurality of different currencies.

The American economy blossomed despite the lack of a single currency. Indeed, the success of the dollar owes a great deal to the process of gradual evolution over which it emerged. The dollar was not established by the diktats of the political élite but gained slow acceptance across the population as a whole.

In contrast, we see the Soviet-style attempts by the central planners of the European Union to impose the euro on a wide diversity of countries, far more diverse than the individual states of America. It has led to economic stagnation and the rise of far-right nationalist parties across Europe. All the core members of the European Union had strong historical links with fascism, and a powerful driving force behind its inception was to prevent a repeat of history. Yet its institutional structures are themselves creating the risk that it will happen again.

Much more generally, it is innovation, evolution and competition which are the hallmarks of a successful system. This is a fundamental message from Hayek and Schumpeter which shines to us across the decades.

Schumpeter coined the superb phrase 'gales of creative destruction'. He argued that innovation led to such gales that they caused old ideas, technologies, skills and equipment to become obsolete. The question, as Schumpeter saw it, was not 'how capitalism administers existing structures ... [but] how it creates and destroys them'. This creative destruction, he believed, caused continuous progress and improved standards of living for everyone.

The model of evolution and extinction discussed and developed in this book possesses exactly these properties. Recall that in the model at any point in time each particular agent, be it a firm, a species or whatever, has its own level of fitness for survival. Those that fall below a critical value become extinct. By adding together the fitness levels of all the component parts, we can measure the fitness of the system as a whole and how it evolves over time.

When we analyze the connection between the overall fitness of the system and the rate of extinction, we see quite clearly that periods immediately following large extinctions tend to have relatively high overall fitness, as new firms are rushed in to fill the gaps opened up by the eliminations. In other words, extinctions essentially play the role described by Schumpeter in his phrase 'creative destruction'. Weaker firms are eliminated and replaced by firms which, on average, have higher levels of fitness.

Figure 14.1 plots the relationship between the change in overall fitness of the system in our model in the current period from the previous period and the rate of extinction in the previous

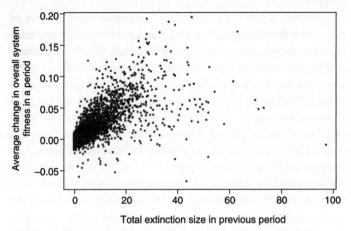

FIGURE 14.1 The change in the overall fitness of the system and the size of extinction in the previous period.

period.* As we can see, there is a strong positive relationship between the two.

We can go on to use the model to illustrate the crucial importance of competition for the fitness of the system as a whole. But we need to make a distinction between competition between firms within an industry and competition between industries – the competitive environment in which the industry operates. This distinction is not merely of theoretical interest. It is one which is rarely distinguished in regulatory practice, with potentially seriously adverse consequences for economic performance.

We can think of the critical level of fitness at which agents become extinct as indicating the general competitiveness of the environment in which they operate. The lower the level of fitness that is needed to survive, the easier is the competitive environment.

* To be precise, the model is solved over 150,000 iterations, and the first 10,000 are discarded. The iterations are then placed into groups of five, in order to compare a reasonable length of 'time' following extinctions with a similar length of 'time' in which they take place, rather than the single iterations of the model solutions. This is to take account of the fact that extinctions often occur in groups, so that several extinction events may follow each other closely. This procedure gives 28,000 observations of periods of extinction rates and system fitness.

229

Until very recently, for example, it was almost impossible for the national airline of a western European country to be forced to cease trading. State subsidies of one form or another, often dressed in the thinnest of disguises, were always available to keep even the most inefficient airlines in business. In the UK, from the end of the Second World War to the 1980s and the Thatcher government, subsidies on a massive scale were paid both to publicly owned industries such as coal and steel and to private ones such as the car industry. More generally, the overall environment in the former Soviet bloc meant that it was much easier for inefficient firms to survive there than it was in the more market-oriented western economies.

We can investigate this in the model by varying the fitness threshold required for the survival of agents and examining the impact on the average fitness of the system as a whole. If we increase the threshold at which agents become extinct, we are making it more difficult for them to survive, and if we reduce it, we are making it easier. We can readily think of this as corresponding to more and less competitive environments, respectively.

Recall from Chapter 11 that in our model an agent is deemed to become extinct if its fitness falls below zero, and its fitness is defined as the sum of the impacts of the strategies of other firms on it. It seems natural to set this extinction level as zero, which we did in the results discussed above. But we can set different levels, so that agents in the model can survive when their overall fitness is negative in the model. It is important to emphasize that this is not violating some natural law. The scales of values within the model are arbitrary, and the only requirement is that they need to be internally consistent with one another.

Figure 14.2 plots the average fitness level of the system as a whole for different values of the fitness threshold. As can be seen quite clearly, the less stringent is the overall environment, the lower is the overall fitness level of the system.* In other words, an

---

* The units of measurement of both the system threshold and system fitness are purely internal to the model and could be calibrated on any scale desired. In other words, the units of measurement themselves have no particular significance, but the result showing the connection between these factors most certainly does.

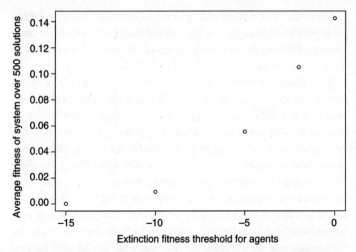

FIGURE 14.2 System fitness and the external environment: system fitness at different thresholds for agent extinction, averaged over 500 solutions of the model.

environment which stresses the need to innovate and to keep pace with the way the world is changing leads to a higher fitness level for the system than one in which an easy life can be had and inefficiency is not punished so ruthlessly. The easier it is for agents to survive, the lower is the fitness of the system as a whole.

So a demanding and competitive external environment is essential for the health of the system as whole. But economic regulators, with the possible exception of the US authorities, are obsessed with something entirely different. Their mission in life is to promote competition *within* the system, to increase the level of competition between the individual agents inside the system.

Schumpeter certainly believed that compelling firms to compete too fiercely drives profit margins down to low levels, so that firms lack the ability, and possibly the incentive, to invest for the longer term. Hayek emphasized the importance of the process of the evolution of strategy, of discovering new and more effective ways of doing things. Both these views imply that strong market positions will be created temporarily for particularly successful innovators. They will be able to exercise market power and raise their profit margins. This is the reward to innovation at the individual level but, as we have seen, this also brings benefits to the system as a whole.

The market power, the monopoly, will not be permanent, because at some unpredictable point a new innovation will spread which will undermine the basis of the existing market power of the monopolist. As long as the institutional rules under which the system operates encourage innovation, we should not worry about market power being exercised by individual firms, for eventually they will be undermined by the process of competition and innovation. Microsoft, for example, may appear to bestride the world at present, but eventually it will undermined by innovation, possibly by the current developments in open system Linux, or more probably by someone who is just now merely thinking of setting up an operation in his or her garage in Silicon Valley. In the meantime, Microsoft is compelled by fear of competition to innovate constantly, bringing benefits to all.

Conventional economics and regulatory authorities do not see the world in this way at all. They want to undermine market power, even when it arises from successful innovation, and seek to create a world in the image of their own theory. A world in the image of the economic textbook model of 'perfect' competition, when all firms offer the same product in a market and none can make more than the absolute minimum level of profit needed to stay alive.

We can use our model to look at this approach to public policy. Recall that pairs of agents in the model can be connected together in different ways. In particular, both can have positive impacts on each other's fitness, implying they act in a co-operative way. Alternatively, one or both can have a negative impact on the other's fitness, implying a competitive relationship between the two. We can increase the number of negative connections and reduce the positive ones, and see what happens. In other words, we can examine the consequences of increasing the level of competition between the component agents of the system. These results are plotted in Figure 14.3. The chart requires a certain amount of explanation.

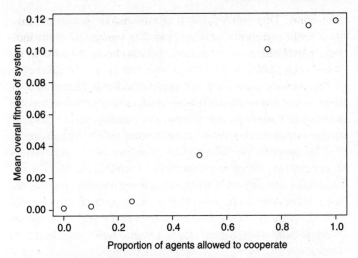

FIGURE 14.3 Overall fitness of the system and the internal level of competition in the system. When the latter is zero, only competitive relationships are permitted between agents. When it equals one, we have the 'normal' operation of the model, with its combination of co-operative and competitive relationships.

The bottom axis shows the proportion of positive connections between agents which are permitted compared to the 'normal' working of the model. When this is equal to 1, model is operating as described in previous chapters, able to replicate the salient facts about biological and firm extinction. It has a mixture of positive and negative connections between each pair of agents. At the value of zero on the bottom axis, no positive connections are permitted.

The implications are clear. Regulation that serves purely to intensify the internal degree of competition in a dynamic, evolving system can have adverse consequences for the system as a whole. As the degree of this kind of regulation increases, initially the adverse consequences are small, but beyond a critical level their impact grows rapidly. And with a high degree of such regulation, the fitness of the system begins to approach zero, at which level it risks total extinction.

In terms of the public policy implications, the authorities need to recognize that within a single industry co-operative relationships between firms are necessary for the fitness of the industry as whole. If competition is enforced too strongly, the overall fitness of the industry is reduced. But protection for the industry as a whole also tends to reduce overall fitness. Firms need an external environment which is competitive but require a certain level of co-operative behaviour between each other within the industry.

These results are so different from standard models of economics precisely for the reasons identified by Schumpeter and Hayek. The model describes a dynamic process of evolution, innovation and competition. In contrast, the world of 'perfect' competition describes a purely static world of equilibrium, one more suited to the socialist central planner than to the real, living world of capitalism.

There is a substantial sector of western society where this is barely understood at all. Governments throughout the west still behave as if the Soviet Union and its Five-Year Plans had been a great success story rather than a catastrophic failure. During the second half of the twentieth century, many people have supported the dramatic expansion of the state on the grounds that this improves social and economic well-being.

Governments now feel obliged to act, or at least to be seen to be acting, on almost every aspect of human existence. As I wrote the first draft of this chapter, for example, the British government announced that it intended to introduce a 'walking strategy'. The aim, apparently, is to advise people to walk. Targets will be set and officials employed to monitor whether or not they have been reached. Exactly how this will be implemented is even now not clear, but the fact that humans have been walking for tens of thousands of years without the benefit of such a strategy is obviously regarded as simply not being good enough. Bicycles have existed for a mere twinkling of an eye in comparison to humanity, but even they are not immune. A new law makes it compulsory for new bicycles to be sold equipped with a bell. Shopkeepers can be fined for failing to comply, and, again, public sector officials will monitor them. However, it will not be illegal for people to ride their bikes without a bell. So officialdom is satisfied if the bike leaves the shop with its sparkling bell, only for it to be removed by the proud new owner immediately. In the leafy part of London where I live, the River Thames winds slowly through acres of trees and pleasant grassland. A wide and well-maintained towpath follows the river. Three years ago, the local council installed wooden chicanes at intervals of several hundred metres in order to, and I am not making this up, 'curb over-enthusiastic joggers'.

The point is not to ridicule such policies, though each has its own particular streak of lunacy. Almost every reader will be able to come up with similar stories. The point is that the state, at both national and local level, now feels obliged to intervene, to have policies regarding almost every aspect of life. Whether walking, jogging or riding a bike, teams of inspectors are on hand to endeavour to ensure that targets are met on these various means of locomotion. Indeed, the current British government has announced no fewer than 8,000 numerical targets since coming to power in 1997.

The urge to intervene, to be seen to be doing something, has reached epidemic proportions. In the British Parliament, for example, there is a procedural device called an Early Day Motion. An MP can put one forward and try to attract support by getting other members to sign the motion. On a single day chosen at random in 2004, the British government was urged in Early Day Motions to: hold a full inquiry into political opinion polls; to give air quality a higher priority; take firm action against 'disablism'; give a posthumous VC to Lieutenant-Colonel Paddy Mayne; introduce Northern Ireland-wide standards for care and access to arthritis treatments; press for the introduction of regulations to improve safety standards in European holiday resorts; increase the amount of funding to hospices; not bring back the poll tax; ensure that members of the British Diplomatic Corps can work safely in Bangladesh; deal firmly with attacks on NHS staff; propose the suspension of the EU–Israel Association Agreement; set up an independent public inquiry into Gulf War Syndrome; support the Pay Up For Pensions march; invest in the East Coast mainline railway;

ban smoking in public; make clear the cost to Oxfordshire County Council of an asylum centre; support small business in legal action against large foreign multinationals; apologize for claiming Iraq had WMDs; amend the finance bill to allow people to invest in films; and, finally, abolish the need to reballot to maintain trade union political funds.

But on very fundamental aspects of social and economic well-being, this massive increase in the role of the state has failed to deliver. Forty or fifty years ago, for example, the typical unemployment rate in countries such as France and Germany was below 2 per cent. The rate fluctuated from year to year, but on average it was very low. Now, the rate is nearly 10 per cent in these economies. In the UK, around 1960 unemployment was typically around 2 per cent also. Now, it is around double that rate, despite a decade of almost continuous falls from the double-digit rates of the early 1990s. Of the major economies only in the US is the unemployment rate now similar to what it was forty years ago.

Crime is now much higher than it was in the decade or so immediately following the Second World War. There are, of course, formidable problems in comparing crime rates over long periods of time, even within the same country. Methods of data collection change and perceptions of what constitutes a crime worth reporting change, so that the data are never absolutely comparable in different years. But it is hard to escape the conclusion that crime is now much higher than it was forty or fifty years ago. In the US, for example, the crime rate, the number of crimes per 100,000 of the population, tripled between 1960 and 1980. It subsequently fell sharply, but remains more than twice as high as it was in 1960. The experience in Britain is qualitatively similar to that of the US, though Interpol estimates suggest that it is now two and half times higher in the UK than it is in the US. The major countries of continental Europe have also seen sharp increases in crime. As late as 1980, the rates in France and Germany were somewhat below those in the US. Now, they are almost twice as high.

Social mobility appears to have declined in many western countries since around 1970. The concept measures the extent to which children end up as adults in the same socio-economic grouping as their parents. The promotion of social mobility has been a prize aim of the social-democratic mode of thinking. Yet, despite an unprecedentedly large amount of public-sector activity, which should at least in principle enhance public well-being, it has declined. The chances of a child born in deprived circumstances remaining poor are higher than they were forty years ago.

We saw in Chapter 4 that residential segregation along racial lines has hardly been eroded at all in the US, despite the intense efforts of public policy to achieve this aim. And income inequality has increased within the capitalist economies over the past thirty years.

In 1900, it was not obvious that the market-oriented industrial economies ('capitalism' for short) which had developed during the nineteenth century would go on to be so successful. As it happens, during the nineteenth century, workers had become considerably better off under capitalism. Real wages had risen, calorific intake had increased and life expectancy had also risen. But absolute living standards were still very low by present-day standards, and work was long and hard. Further, income inequality was very much more evident than today – the gap between the working-class/lower-middle-class majority and a fairly small minority was very marked.

In these circumstances, it was perfectly reasonable to consider alternative forms of economic organization. In the alternative approach which came to dominate, the state played a central role. There were further increases in living standards during the first half of the twentieth century. This is an important reason why, after 1917, there were no attempts to overthrow an established capitalist system that were even remotely successful. An additional reason for this was the marked narrowing in the degree of inequality which took place.

But the first half of the century was marred by the high western death rates in the First World War, and the high unemployment rates of the inter-war years. So although revolutionary overthrow disappeared as an alternative in the west, attachment to a more planned economy in the which the state played a substantial role remained strong.

This was the model which all western governments were obliged to accept as part of the post-Second World War settlement. Everywhere, including the US, the claims of the state on national output and its role in social and economic policy were very much bigger in the second than in the first half of the twentieth century.

We have now had almost sixty years in which to evaluate this model – the 'social democratic model' for short. The model is simply a variant, or a development, of an intellectual tradition that was first articulated in the nineteenth century. But does the word 'progressive' in the twenty-first century still necessarily mean attachment to a large role for the state? This question perhaps needs a bit more precision. The state is obviously very important in setting the rules of the game. The question really is: does 'progressive' continue to mean a belief in a large role for the state as player, in addition to its role as arbiter/referee?

In the second half of the twentieth century, the strength of the capitalist mode of production became absolutely apparent. Living standards reached hitherto unimagined levels. Life expectancy rose dramatically. Countries outside the west, some with large populations, which had previously lived close to starvation levels also shared in the increase in prosperity. Acute poverty and starvation became the characteristics of anti-capitalist regimes.

But it is not the *size* of the state as such which has brought this about. Different western countries have experienced different sizes of state intervention in the economy, and there is no obvious relationship between this and economic performance. And, as we have seen, the period in which the state has seen a massive increase in its importance in western society has also been the period in which most countries moved away from rather than towards the outcomes that the social democratic model promised. Unemployment is up. Crime has increased. Income inequality has widened. And social mobility has fallen.

This is not to say that all public policy necessarily ends in failure. Almost at random, some will succeed. But the way of thinking, the

Weltanschauung, which relies upon, which believes in detailed planning to achieved precise, carefully monitored aims is inherently doomed to fail.

In Schelling's model of segregation, it is impossible for any individual to plan a strategy which will guarantee that his or her aims are met. In the Prisoner's Dilemma or in chess we rarely know what the best strategy is at any point in time. In the model of extinction, it is not feasible to plan a strategy which will increase the probability of survival to any substantial extent. These games, these models, are simple. Their rules are straightforward and easily understood. But they give rise to complex interactions between agents. This in turn creates such uncertainty about the future consequences of actions that the ability to plan the actions in great detail and to bring about their intended consequences is very limited.

Administering public policy is a far more complicated task, and the dimensions of the problem are even greater than those described by our models and games. Not surprisingly, the cast of mind which believes in detailed planning and monitoring usually fails.

In many ways, companies understand this intuitively. The potential rewards from gaining even a very partial insight into the future are substantial. But no one can rely upon this as a basis for success. Company brands fail for a whole variety of reasons. After the event, after the failure, like any good historian or social scientist, the brand manager or ad agency responsible for the account can construct a narrative to account for failure. But these stories, to give them an equally accurate but more homely description, are precisely that. We can never know for sure why a brand or even a company failed. More importantly, we can never know for sure in advance how a new product or company will perform.

Here, for example, is perhaps the most spectacular brand failure in the whole of the twentieth century: in the early 1980s, Coca-Cola's leading position in the soft-drinks market was gradually being undermined by Pepsi. The latter brand had built successfully on its 'Pepsi Challenge' campaign, a blind test for consumers on its own product and Coca-Cola. On taste, Pepsi seemed to be winning hands down. After a massive research effort, Coca-Cola responded by withdrawing its own product and introducing New Coke on 23 April 1985. On 11 July 1985, New Coke itself was withdrawn and the old brand reintroduced because sales had collapsed.

And here is the CEO of Coca-Cola, Donald Keough, justifying this decision at the press conference: 'The simple fact is that all the time and money and skill poured into consumer research on the new Coca-Cola could not measure or reveal the deep and abiding emotional attachment to original Coca-Cola felt by so many people. The passion for the original was something that caught us by surprise. It is a wonderful American mystery, a lovely American enigma, and you cannot measure it any more than you can measure love, pride or patriotism.'

Almost all brands fail eventually. More importantly, most fail very soon after their introduction. The attrition rate in early life is very high. Companies know this to be an inherent fact of life and they respond by constant innovation, constant testing of new ideas and new brands, new products. They do so within a frame-work, an institutional structure, which itself has adapted over time, has evolved, and which facilitates flexibility and innovation. New Coke was indeed a spectacular failure, but the Coca-Cola board did not sit around and commission endless, meaningless reports into how they would guarantee to avoid such failure in future. Still less did they try to pretend that somehow the great Five-Year Plan was still being fulfilled. Instead, they acted.

Karl Marx famously wrote that the motto of capitalists was 'Accumulate, accumulate, that is the law of Moses and the Prophets!' As in many other respects, Marx was completely wrong. 'Innovate, innovate!' – that is the guiding principle which companies have used to try to overcome the inherent and pervasive uncertainty which surrounds all their decisions. It is the best strategy for indi-vidual survival, and it is a strategy from which we all, as consumers and citizens, have benefited immensely.

Suggestions for Further Reading

This is not intended to be a comprehensive set of references covering every single point made in the text. Rather, it is a mixture, documenting in more detail some of the articles and books mentioned in the main body of the book, and giving suggestions for further reading. These suggestions below can be easily augmented by a search on the World Wide Web, either by topic or by author.

By the very nature of the subject, a substantial part of the material is not readily accessible by readers without some familiarity with mathematics. But the level of difficulty varies, and the fact that part of an article contains maths should not always act as a deterrent.

A very good survey of modern mathematical modelling of evolution and extinction is given in Barbra Drossel's article 'Biological Evolution and Statistical Physics' in *Advances in Physics*, Vol. 50, pp. 209–95 (2001). This contains the references to many relevant articles, such as the Newman and Solé Manrubia models of extinction, and empirical work on the relationship between the size and frequency of extinctions in the fossil record. These are all in varying degrees rather technical. The extinction models are described in M. E. J. Newman, 'A Model of Mass Extinction', *Journal of Theoretical Biology*, Vol. 189, pp. 235–52 (1997), and in R. V. Solé and S.C. Manrubia, 'Extinction and Self-Organized Criticality in a Model of Large-scale Evolution', *Physical Review E*, Vol. 54, R42–R45 (1996).

My own web page, http://www.paulormerod.com, has a number of articles that I have drawn upon for material in this book. Even when writing for specialist academic journals, I do try to explain things as far as possible in English, but there is often an inevitable level of technicality. The relationship between the size and frequency of extinctions of firms in the US, written with Will Cook, is published as 'Power Law Distribution of the Frequency of Demise of US Firms' in *Physica A*, Vol. 324, pp. 207–12 (2003). A similar article looking at evidence from eight other OECD countries, written with Corrado de Guilmi and Mauro Gallegati, 'Scaling Invariant Distributions of Firms' Exit in OECD Countries', is also in *Physica A*, Vol. 334, pp. 267–73 (2004).

A paper on my website written entirely in English addresses the question of how much learning it is feasible for economic agents, such as individuals and firms, to actually carry out. Entitled 'What Can Agents Learn?', it was given as the opening plenary address to the Australian Economic Society conference in Canberra in September 2003.

An excellent example of how to convey scientific material in good English to a non-specialist audience is the works of Stephen Jay Gould. He has written on a wide range of important matters in biology. Perhaps the most relevant for the material in this book is *Wonderful Life* (Penguin, 1991), which describes the explosion in different forms of life which took place in a biologically very short space of time in the Cambrian era some 550 million years ago.

The experience of firms in the period around 1900, when so much innovation took place, is described and analyzed in Alfred Chandler's masterpiece, *Scale and Scope: the Dynamics of Industrial Capitalism* (Harvard University Press, 1990). The survival and extinction of large firms during the twentieth century is documented in Neil Fligstein's book *The Transformation of Corporate Control* (Harvard University Press, 1990) and Leslie Hannah's article 'Marshall's "Trees" and the Global "Forest": Were "Giant Redwoods" Different?' in N. R. Lamoreaux, D. M. G. Raff and P. Temin (Eds), *Learning by Doing in Markets, Firms and Countries* (National Bureau of Economic Research, 1999).

Two very entertaining books about what really goes on inside giant corporations are *Barbarians Led by Bill Gates* by Marlin Eller and Jennifer Edstrom (New York, Henry Holt, 1998), its title no doubt inspired by the story of the RJR Nabisco corporate buyout, and *Barbarians at the Gate: The Fall of RJR Nabisco* by Bryan Burrough and John Helyar (Harper Business, 1991). Although covering a slightly different area of economics than this book, P. J. O'Rourke's *Eat the Rich: A Treatise on Economics* (Atlantic Monthly Press, 1998), gives a humorous analysis of economic growth which tells us at least as much as the vast literature in academic economics about this very important phenomenon.

An immense amount of historical data about the world economy is available in the works of Angus Maddison; one that is used widely is *Monitoring the World Economy 1820–1992* (OECD, Paris, 1995). Robin Marris has an interesting account of world poverty and inequality in his short book *Ending Poverty* (Thames and Hudson, 1999).

The Schelling model was years ahead of its time in the methodology which was used, one which is now being used successfully in a rapidly increasing number of areas in economics and sociology. The original reference is 'Dynamic Models of Segregation', *Journal of Mathematical Sociology*, Vol. 1, pp. 143–86 (1971).

A search on the web will yield huge numbers of pages on popular games such as the Prisoner's Dilemma. Some will even offer the opportunity for interactive play. A simple interactive version of Hotelling's location game described in Chapter 7 can be found on Volterra Consulting's website, http://www.volterra.co.uk.

A seminal reference on co-operation in the Prisoner's Dilemma is Robert Axelrod's *The Evolution of Co-operation* (New York, Basic Books, 1984), and a more recent work is his *The Complexity of Cooperation: Agent-Based Models of Competition and Collaboration* (Princeton University Press, 1997). At the time of writing, a good web page containing much recent material on this game and on the impact of uncertainty is http://plato.stanford.edu/entries/prisoner-dilemma.

A very thoughtful discussion of the role of game theory in economics is David Kreps's *Game Theory and Economic Modelling: Clarendon Lectures in Economics* (Oxford University Press, 1991).

Written by a game-theory enthusiast, the book nevertheless is careful to discuss its limitations. A detailed and fascinating account of the development of game theory in economics is Philip Mirowski's book *Machine Dreams* (Cambridge University Press, 2002).

A lot of material on power laws can be found on the world econophysics website, http://www.unifr.ch/econophysics, where many papers are posted prior to publication. Examples of such papers include L. A. N Amaral, S. V. Buldyrev, S. Havlin, H. Leschorn, P. Maass, M. A. Salinger, H. E. Stanley and M. H. R. Stanley 'Scaling Behaviour in Economics: I. Empirical Results for Company Growth', *Journal of Physics I France*, Vol. 7, pp. 621–33 (1997); R. L. Axtell, 'Zipf Distribution of US Firm Sizes', *Science*, 7 September 2001; A-L. Barab'asi, R. Albert and H. Jeong, 'Scale-Free Characteristics of Random Networks: The Topology of the World Wide Web', *Physica A*, Vol. 281, pp. 69–77 (2000); F. Liljeros, C. R. Edling, L. A. N. Amaral, H. E. Stanley and Y. Aberg, 'The Web of Human Sexual Contacts', *Nature*, Vol. 411, pp. 907–8 (2001). An introduction to this type of analysis in financial markets is J.-P. Bouchard and M. Potters, *Theory of Financial Risks: from Statistical Physics to Risk Management* (Cambridge University Press, 2000).

There is a rapidly growing literature on percolation across different kinds of networks, much of which is in the econophysics literature and can be found on the website mentioned above. An important early – early in the context of this field – contribution is by Duncan Watts and Steve Strogatz, 'Collective Dynamics of "Small World" Networks', *Nature*, Vol. 393, 4 June 1998. For an unusual and much more qualitative example, see my paper with mediaeval historian Andrew Roach, 'The Mediaeval Inquisition: Scale-free Networks and the Suppression of Heresy', *Physica A*, vol. 339, pp. 645–52 (2004).

Important criticisms of general-equilibrium theory from within mainstream economics itself are R. Radner, 'Competitive Equilibrium Under Uncertainty', *Econometrica*, Vol. 36, pp. 31–58 (1968); H. Sonnenschein, 'Market Excess Demand Functions', *Econometrica*, Vol. 40, pp. 549–63 (1972); C. J. Bliss, *Capital Theory and the Distribution of Income* (Elsevier North-Holland, 1975).

The 2001 economics Nobel prize-winners, George Akerlof and Joe Stiglitz, were instrumental in introducing the concept of bounded rationality into economics. The seminal article is probably Akerlof's 'The Market for "Lemons": Quality Uncertainty and the Market Mechanism', *Quarterly Journal of Economics*, Vol. 84, pp. 488–500 (1970), and an example of Stiglitz's work in this area is his article with Steven Salop, 'Bargains and Ripoffs: A Model of Monopolistically Competitive Price Dispersion', *Review of Economic Studies*, Vol. 44, pp. 493–510 (1977).

The 2002 Nobel prize-winners, Vernon Smith and Daniel Kahneman, have taken economics even further by building up impressive amounts of evidence which show that agents in general do not behave as they are assumed to do in mainstream economic theory. Their Nobel prize lectures were published in the *American Economic Review* in 2003. Smith's is entitled 'Constructivist and Ecological Rationality in Economics', *American Economic Review*, Vol. 93, pp. 465–508, and Kahneman's is 'Maps of Bounded Rationality: Psychology for Behavioral Economics', *American Economic Review*, Vol. 93, pp. 1449–75.

Successful theoretical models based on the assumption that, relative to the complexity of the task, agents have low levels of cognition are being built. These give better accounts of important empirical phenomena than do models built on the conventional economic assumption that agents have high levels of cognition. Examples include Duncan Watts, 'A Simple Model of Global Cascades on Random Networks', *Proceedings of the National Academy of Science*, 99, 5766–5771, 2002, my own article 'Information Cascades and the Distribution of Economic Recessions in Capitalist Economies', *Physica A*, 342, 556–568, 2004, and Szaboks Mike and J Doyne Farmer, 'An Empirical Behavioral Model of Price Formation', posted on the Los Alamos non-linear website, arXiv:physics/0509194 v2.

Both Hayek and Schumpeter wrote almost exclusively in words rather than maths. Both, and Hayek in particular, were very productive authors. Perhaps their single most famous works are Schumpeter's *Capitalism, Socialism and Democracy*, published in

1942, and Hayek's phenomenal bestseller *The Road to Serfdom*, published in 1944. There are numerous reprints of both. The University of Chicago reprinted the Hayek book on its fiftieth anniversary in 1994, and there is a Perennial Press 1962 edition of the Schumpeter volume.

Index